OMES'

ROYAL WARRANT

NG MACHINERY

OBTAINABLE

Ransomes, Sims & Jefferies

Agricultural Engineers

RANSOMES'

LAWN MOWERS
THE BEST IN THE WORLD.

Ransomes, Sims & Jefferies
Agricultural Engineers

A history of their products

Brian Bell MBE

First published 2001, reprinted 2008

ISBN 978-1-903366-15-8

A catalogue record for this book is available from the British Library

Published by

Published by
Old Pond Publishing
Dencora Business Centre
36 White House Road
Ipswich IP1 5LT
United Kingdom

www.oldpond.com

Edited by Julanne Arnold
Cover design and book layout by Liz Whatling
Printed and bound in China

Contents

Preface

This book takes a broad view of two hundred years of Ransomes' products made at Ipswich and used throughout the world. Their vast range includes ploughs and combine harvesters, steam engines and thrashers, lawn mowers and electric trucks, motor cultivators and crop sprayers, as well as many less familiar machines.

Robert Ransome opened an ironmongery business in Norwich in 1774 and then moved to Ipswich in 1789. Before long Ransomes' ploughs were world famous and at the height of the industrial revolution the Ipswich company was one of the largest manufacturers of farm machinery in the world.

Many medals were awarded for mechanical excellence and thousands of prizes have been, and still are, won with Ransomes' ploughs at National and World Ploughing Championships.

There have been many pioneering achievements. Robert Ransome invented the self-sharpening ploughshare before Nelson was victorious at Trafalgar; he made his first lawn mower in 1832 and the company was making potato planters, maize shellers and swath turners before the turn of the century. James Edward Ransome made his first motor mower in 1902; the first electric vehicle was tested at the Orwell Works in 1915 and Ransomes, Sims & Jefferies were making electric trolleybuses in 1924.

The scope of the business was widened by the acquisition of other companies such as Howards of Bedford, Catchpole Engineering, Johnson's Engineering and the Dorman Sprayer Co.

In a relatively short book it is impossible to give the full story of Ransomes' history. Anyone who has visited the archive at the Museum of English Rural Life, Reading, will realise what a wealth of information remains to be researched.

However, I hope that in the following pages the reader will find an accurate introduction to the range of familiar and unfamiliar machines and implements made by Ransomes at Ipswich since 1789.

I have brought the story up to date with the sale of Ransomes' agricultural division to the Electrolux group in 1987 and of the grass machinery division to the American Textron Corporation in 1998. The Ransomes name still appears today on turf maintenance equipment made in Suffolk and exported worldwide.

Brian Bell

February 2001

Acknowledgements

I am indebted to many people, and in particular those who spent much or all of their working lives at Ransomes, for their help in compiling this book. Rick Barnes, Doug Cotton, Perry Crewdson, Roger Cutting, Bruce Dawson, Fred Dyer, Stuart Gibbard, Mark Grimwade, Charles Halliday, Bill Love, Bob Malster, Bob Mee, Sid Palmer, Stephen Smith, Geoff Teague and Len Whurr have provided me with a wealth of information and anecdotes about life at the Orwell and Nacton Works.

Special thanks are due to Textron Golf Turf & Specialty Products, and in particular Selina Flynn and Bob Rendle for their support and to the Museum of English Rural Life, Reading for permission to reproduce illustrations, and particularly to Dr. Jonathan Brown.

Chapter 1

Ransomes of Ipswich

Robert Ransome, son of Quaker schoolmaster Richard Ransome, was born at Wells in Norfolk in 1753. Following an apprenticeship with a Norwich ironmonger, he opened his own ironmongery shop at No. 50 Market Place, Norwich in 1774. His next enterprise was the manufacture of cast-iron plough shares at a small iron foundry near White Friars Bridge in Norwich. Production of brass castings was added a year or so later and at that time the White Friars foundry and another in Cambridge were the only iron foundries in East Anglia. Robert Ransome adopted the name of Ransome & Co in 1784 and in the following year he patented a process for tempering cast-iron plough shares. The shares were advertised in the *Norwich Mercury* newspaper in March 1785. They were offered for sale at Ransomes' ironmongery warehouse in the Market Place, at the White Friars Bridge foundry, at all the ironmongers in Norwich and at about fifty towns throughout the eastern counties.

Robert Ransome moved his business to the Suffolk town of Ipswich in 1789 with £200 capital and one employee. One of the country's oldest towns, founded around 600AD at the lowest bridging of the river Orwell, Ipswich offered a safe haven and port facilities for the import of raw materials. A new iron foundry was set up in a disused maltings and the plough shop, which backed on to an outdoor skittle alley, was situated in St Margaret's Ditches, now called Old Foundry Road. The production of Robert Ransome's self-sharpening chilled cast-iron plough share, made by a process allegedly discovered by accident and patented in 1803, was an important event in the early days of the business, and before long self-sharpening shares were out-selling the earlier wrought-iron ones.

Robert Ransome made an increasing number of ploughs and was winning awards for his new ideas when in 1908 he registered a patent for the standardisation of plough parts. This major development in engineering production meant that a spare part for a plough or other implement could be taken off the shelf without the need to have a replacement part hand-made by a local blacksmith.

His son James, following an apprenticeship with his father, opened a new foundry with another former Ransomes' apprentice at Great Yarmouth. James moved back to Ipswich in 1809 and joined his father's prospering business, which at that point became Ransome & Son. Dealerships to sell Ransomes' implements and self-sharpening plough shares had been established throughout East Anglia and by 1810 Ransomes' ploughs were being exported to South Africa and Canada.

The company already had a warehouse for the display and sale of ploughs in Norwich and another was opened near the Ipswich Corn Exchange in 1821. A period of agricultural depression occurred soon after James joined the business but fortunately Ransome & Son had other engineering interests. The Ransome family was involved in the formation of the Ipswich Gas Supply Company to provide town gas for Ipswich in 1817 and they retained an interest until it became a separate company in 1822. William Cubitt, appointed as Ransome's company engineer in 1812, added general engineering to the business and in 1818 he took charge of a contract to build a cast-iron bridge which spanned the River Gipping for the next 105 years. Cubitt also invented wind-regulated vanes for windmills and a human treadmill. The latter was used to grind corn and one of these treadmills was installed in Ipswich prison on the site of the present County Hall in St Helen's Street. The founder's younger son, also Robert, was apprenticed to his father in 1810 and was made a partner in 1818 when the firm became Ransome & Sons.

Robert Ransome gave up his active role in the business in 1825 and the firm changed its name to J. & R. Ransome. After the founder's death in 1830, his grandson James Allen Ransome became a partner and the business was re-named as J. R. & A. Ransome. James installed a steam engine in the foundry in 1831 and he proved to be a best-selling author with his book *The Implements of Agriculture* published in 1843. Meanwhile, cultivators, land rollers and early designs of threshing machine were added to the Ransomes product range and the first Budding patent lawn mower was made under licence at Ipswich in 1832. The lawn mower was a great labour-saver for the men who were employed to cut grass with a scythe, a time-consuming task that until then had meant that only the very rich could afford the luxury of a neat lawn trimmed by hand. More than 1,500 Budding pattern lawn mowers had been made at Ipswich by 1852.

Charles May, also a Quaker, joined the firm in the early 1830s. He was responsible for the production of railway track components at Ipswich and this became a considerable part of the business during the mid-1800s' railway building boom. May's greatest achievement was a patent design for cast-iron chairs used to secure rails to wooden sleepers.

Robert Ransome, the founder, 1753 - 1830.

The first Royal Agricultural Show was held at Oxford in 1839 and one of the first Gold Medals given by the Royal Agricultural Society of England was awarded to Ransomes for their stand at the show. About six tons of ploughs, chaff cutters, thrashers and other machines were hauled a hundred or more miles from Ipswich to Oxford on horse-drawn wagons.

J. R. & A. Ransome began the transfer of their business to a dockside shipyard on the River Orwell in 1841 and the move to the new Orwell Works was completed with the closure of the St Margaret's Ditches foundry in 1849. The new premises provided the space needed for the thousand or so employees to meet the growing demand for agricultural implements. The railway components were made at a separate site but within eight years this operation too had moved to the Orwell Works. Hand- and steam-driven threshing machines were being made at the Orwell Works. At the 1841 Royal Show Ransomes demonstrated steam threshing with a portable engine, and they claimed another first at the 1842 Royal Show with a self-propelled steam engine.

Plough production steadily increased over the years: the YL plough with steel beams and handles was introduced in 1843 and the company was even making one for elephant draught in the mid-1840s. In recognition of his contribution to the business Charles May was made a partner in 1846 when J. R. & A. Ransome became Ransomes & May. Arthur Biddell, a farmer at Playford near Ipswich, invented a wooden-framed cultivator or scarifier with two rows of tines in the early 1800s. He joined Ransomes in the early 1840s when an improved version of the Biddell scarifier with an iron frame was made at the Orwell Works. Food preparation machinery for livestock was also made at the Orwell Works until 1872 when Reuben Hunt bought the designs and transferred production to his factory at Earls Colne in Essex.

The workforce was well over a thousand in 1844 when Ransomes published their first comprehensive catalogue illustrating a wide range of products including wooden and iron ploughs, rollers and hay rakes, Biddell's scarifier and other field implements. The 1850s were a decade of change. Ransomes opened a London office at Great George Street, Westminster in 1850 and they were a prominent exhibitor at the Great Exhibition at Crystal Palace in 1851 when their display of agricultural products included steam engines, thrashing machines, ploughs, mills and cake crushers. A printing press for an Ipswich printer and equipment for Greenwich Observatory were built at the Orwell Works in the 1850s. Ransomes' export trade grew at a rapid rate partly helped by emigrants, who took Ransomes implements with them to help start their new lives in the Colonies. The company even tried their hand at shipbuilding, launching a steam-powered iron ship called *The Chevalier* on the River Orwell in 1851, but it was probably the only one of its kind.

Charles May left the business in 1851 to become consulting engineer and later teamed up with a Mr Brown to form Brown & May at Devizes. James Allen Ransome invited his nephew William Sims, who had served his apprenticeship with the company, to become his partner and, trading as Ransomes & Sims, they opened a new retail Town Warehouse in the Princes Street area of Ipswich in 1854. The Orwell Works surrounded a small bell foundry, owned by Billy Bowell where one-off castings were sometimes made for Ransomes' prototype designs. Ransomes & Sims occasionally made large castings for the bell founder and one of the bells in Tattingstone Church, south of Ipswich, bears the inscription 'Cast by Ransomes & Sims 1853'. Ransomes bought the bell foundry premises in 1941 following the death of the owner.

The 1850s and 1860s saw the steam engine and agricultural sides of Ransomes & Sims' business grow at the expense of their other interests. A joint project with John Fowler resulted in the production of the first steam balance plough which was tested at Nacton in 1856. Ploughing with portables and later with traction engines was coming into fashion by the early 1860s and Fowler established his own factory at Leeds. John Head, who was apprenticed at the Orwell Works in 1848, became a leading light in the development of steam engines and this was recognised when he became a partner in Ransomes, Sims & Head in 1869. Extra space was provided for steam engine and thrashing machine production by extending the dockside works in 1871. Mainly due to their other engineering interests, sales of traction engines in Britain were well below those of their major competitors including Burrells of Thetford and Garretts of Leiston. However Ransomes enjoyed considerable success in the export market, selling hundreds of steam engines to many countries including Russia, Argentina, Turkey and South Africa.

Manufacture of railway components was transferred to Ransomes & Rapier, a new Ipswich company formed in 1869 by James Allen Ransome, R.J. Ransome and R.C. Rapier. Both James Allen, the senior partner in Ransomes' agricultural machinery business, and his son Robert were also partners in Ransomes & Rapier. The new company remained in business until 1988 and among the products they made during their long history were cranes and massive walking draglines for open-cast coal mines.

Allen Ransome, another of J. A. Ransome's sons, founded a woodworking machinery business trading as A. Ransome & Co. at Chelsea in 1868. The company moved to Newark in Nottinghamshire in 1900 where they made machinery for manufacturing wooden sleepers and railway carriages. The Newark company widened its interests with the production of ball bearings, some of which were used for their woodworking machines while others were sold to motor car manufacturers. The ball bearing business thrived and a new company called Ransomes & Marles was established in 1917, which in 1969 joined forces with Hoffman & Pollard to form Ransomes, Hoffman & Pollard.

The Kelly's 1883 *Directory for Suffolk* recorded that agricultural implements made by Ransomes, Head & Jefferies and by E. R. & F. Turner of St Peters Works were one of the main exports from Ipswich. It noted that Ransomes' Orwell Works, with 900ft frontage at the east end of the dock, was well placed for the export trade. About twelve acres of land almost entirely roofed over at the Orwell Works provided employment for 1,400 men and boys who were engaged in the manufacture of portable steam engines, thrashing machines, horse rakes and ploughs exported in large numbers to all parts of the world. The company's London office was in Gracechurch Street, and there were showrooms in Edinburgh and Norwich. Closer to home, Ransomes' warehouses and showrooms were located in Princes Street and Queen Street, Ipswich and at Risbygate Street in Bury St Edmunds.

John Jefferies, who was apprenticed to Ransomes & Sims in 1856, progressed to the position of export manager and married James Allen Ransome's daughter. He was made a partner in 1880 and the company became Ransomes, Head & Jefferies in 1881. Following the death of John Head, former partner

William Sims renewed his interest in the business and it became a private company in 1884, trading as Ransomes, Sims & Jefferies Ltd, with John Jefferies as managing director. It became a public limited company in 1911.

Non-original spare parts for Ransomes' implements were already being made by other companies in the second half of the nineteenth century and in an attempt to stop this trade, Ransomes, Head & Jefferies took the matter to the High Court in 1882. The company obtained an order restraining the defendants and any other manufacturer from making any ploughs or plough parts marked with any registered Ransomes trademark. The case lasted for eleven days and about a hundred witnesses either gave evidence or made affidavits on the matter.

In 1892 Ransomes introduced a number of new machines including the 'Leviathan' thrasher, 'Ipswich' and 'Orwell' cultivators, a maize sheller, the Jarmain patent swath turner, a disc harrow and a potato planter. The company were granted their first Royal warrant in 1901 and made the world's first petrol-engined lawn mower in 1902. A new plough works was opened in 1903 and Mr J.E. Ransome introduced a wheeled tractor in the same year. Local farmers were given a demonstration of the prototype tractor with a four-cylinder 20hp petrol engine pulling a three-furrow riding plough. Trading had continued at the Norwich warehouse when Robert Ransome moved to Suffolk but it was closed in 1909 and an agent was appointed to sell Ransomes' machinery to Norfolk farmers. Other depots were opened from time to time at Liverpool, Birmingham and Hull but those at Ipswich and Bury St Edmunds were the only ones to remain open until the late 1950s.

The Orwell Works occupied twenty-five acres in 1911 when the production area was enlarged with a new a thrasher store in Cavendish Street known as the White City and a new pattern shop was built over the quay. There were 2,500 employees at the time and Ransomes' products were being exported to all parts of the world. The onset of the First World War saw a large part of the works turned over to armament production. Farm implements were still made but the works also built 790 FE2b biplane fighters as well as 650 canvas-covered steel-framed aeroplane and airship sheds at the White City. Other departments were engaged in the production of various items including 50,000 ammunition boxes and many thousands of bombs and shells.

A great deal of Ransomes' success in the years before the Great War was based on their export business but much of this trade had disappeared and some was permanently lost when hostilities ended in 1918.

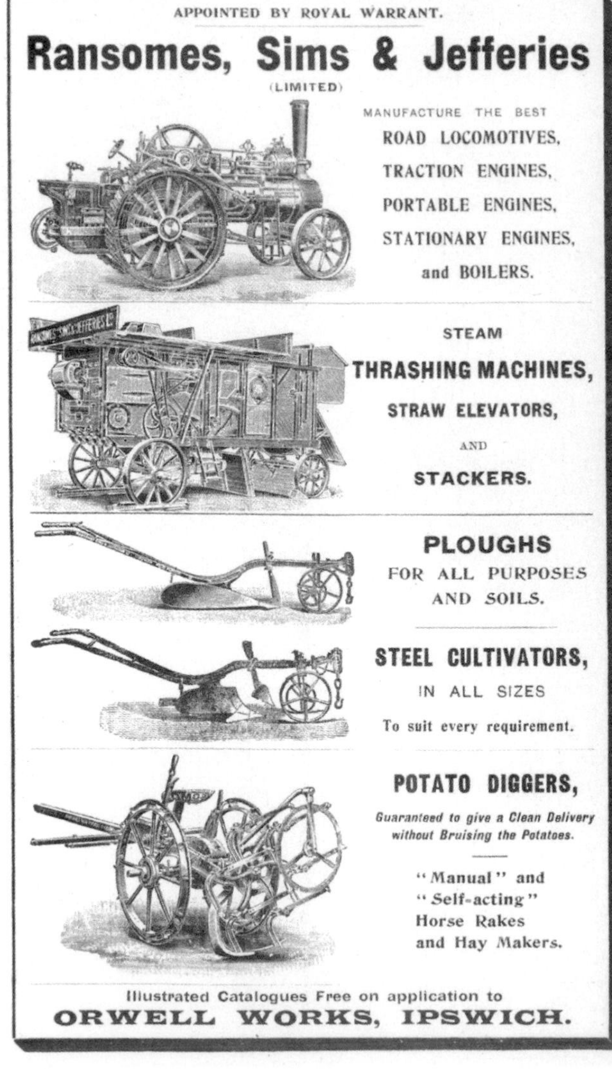

Published in 1913, this advertisement shows the wide range of products made at the Orwell Works.

With the return of peace Ransomes bought their first company motor car, a 15hp Napier Landaulet, and although they enjoyed a short period of buoyant home sales, within three years they faced an agricultural slump and had to lay off 1,500 men.

Ransomes formed an association with Ruston & Hornsby Ltd of Lincoln in 1919. The original Hornsby business was established at Grantham in 1815 and by 1851 about 400 people were employed there. The companies retained their separate identities but they standardised their range of products in order to avoid duplication and to benefit from mass-production techniques. The association, which remained in place until 1940, gave Ransomes access to a wider range of agricultural machinery including reapers, mowers and hay rakes.

A battery-electric vehicle was tested at the Orwell Works in 1915 and Britain's first battery-powered electric truck was made at Ipswich in 1918. Ipswich was one of the first towns to replace its trams with the more versatile electric trolleybuses and to this end Ransomes made their first single-deck trolleybuses in 1924. Double-deck trolleybuses were added in the early 1930s and some of them remained in service on the town's streets until 1963. Ransomes made their first mains-electric lawn mower in 1926 and their first reach truck appeared in 1927.

Meanwhile, developments in Ransomes' farm machinery included the first self-lift plough in 1919, the Weetrac tractor-mounted plough with a mechanical lift in 1927 and the introduction of grass- and grain-drying equipment in 1933. The acquisition of James & Frederick Howard Ltd of Bedford in 1932 widened Ransomes' range. This company had been making all kinds of farm implements including ploughs, cultivators, drills, horse hoes, binders and

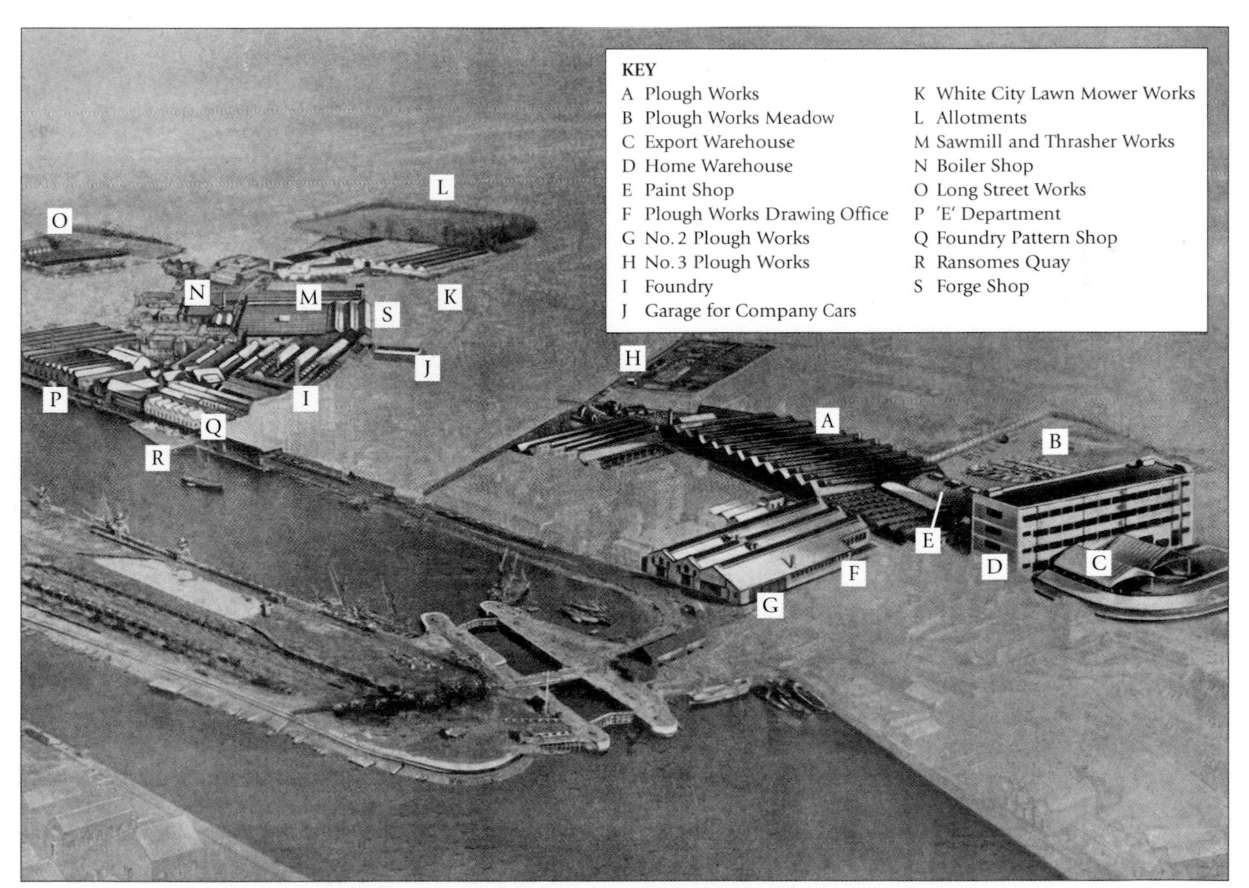

The Orwell Works in 1933.

hay machines since 1811 and had gained a very high reputation for the excellence of its products. Ransomes, Sims & Jefferies' catalogue of Howard ploughs and implements for 1932 explained that as they had taken over the manufacture of Howard agricultural machines they alone were in a position to supply genuine spare parts. Customers were urged to make sure that all spares carried one of the registered Howard trademarks 'in order to ensure continuance of the good results achieved in the past'. To emphasise the point, a footnote in the catalogue pointed out that proceedings had been taken against sixty-six iron founders and others for unlawfully using a registered Howard trademark and that penalties had varied from £5 to £400.

Ransomes' branches were opened in the Argentine

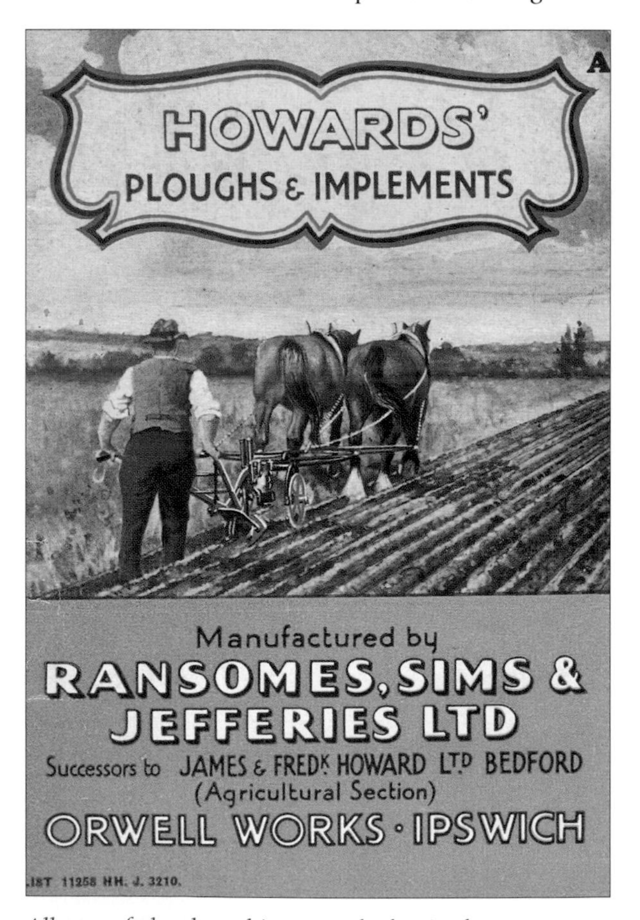

All sorts of ploughs, cultivators and other implements were added to the Ransomes product range when they acquired J. & F. Howard of Bedford.

Crane engine No. 31066 on the Orwell Works quay in 1933 where it was used to load and shunt railway wagons.

and at Cape Town and Durban in South Africa in the 1930s when the Ipswich retail branch was also selling Rushton tractors and other makes of farm machinery including Bamford, Massey-Harris and International Harvester. The thirties was a period of agricultural depression and to compensate for the serious downturn in sales of farm machinery Ransomes obtained various engineering contracts including the manufacture of metal-turning lathes for Elliotts.

The retail business moved in 1935 to a new showroom built on the site where Robert Ransome's private residence once stood in Princes Street. Ransomes were also stockists of milking equipment and they sold vast quantities of United Glass Blowers' milk bottles to Suffolk dairy farmers. Retail branch customers paid their accounts once a year in those days; bills were sent out in December and farmers were invited to the branch on market days in January. Having settled the bill, customers were entertained to a cold lunch with helpings of roast beef carved by a senior company director. The Princes Street branch

Schiffner y Cia were sole agents for Ransomes, Sims & Jefferies machinery in Buenos Aires.

was taken over by Mann Egerton in the mid-1950s and Ransomes' retail warehouse at Bury St Edmunds was closed in 1958.

Sports activities date back at least to 1845 when a fête and cricket match were organised in Christchurch Park for Ransomes' workmen and their wives. Rugby, cricket and athletics clubs were established in the mid-1870s and a football club was added in 1892. There was great rivalry among the departmental tug-of-war teams and the Parkside 'Heavy' team had quite a reputation. Various sports fields were used until 1902 when the Orwell Works sports ground was opened in Sidegate Lane. A fine new pavilion with a large hall, billiards room, bar and changing rooms complete with baths was built there in 1936, and membership of the Orwell Works Sports Club, which included soccer, hockey, athletics, darts and table tennis, cost 1d a week. There was a small extra charge for the tennis club and for the rowing club with a boathouse on the River Orwell. The Athletics and Sports Club thrived for a hundred years or so but reduced staff numbers meant a gradual reduction in its activities.

THE OLD SICK FUND

As early as 1817 Ransomes established a society, later to become known as The Old Sick Fund, to give help to employees unable to work through sickness or injury.

Thirty years later they provided a dining hall at the Orwell works for those workmen who lived too far away to go home for meals. Evening meetings of the Mental Improvement Society were also held in the hall. Membership was 1d a week but less than 25 per cent of the Ransomes workforce attended the classes designed to help them improve their reading and writing.

In 1939 Ransomes, Sims & Jefferies celebrated their 150th anniversary at Ipswich but the outbreak of war meant a return to armament production. The Orwell Works made 17-pounder gun carriages, components for the Merlin engine used for Spitfire and Hurricane fighter planes, bomb trolleys, parts for Crusader tanks and other items to help the war effort. Under the

A boat was needed for a visit the Ransomes retail branch in Princes Street, Ipswich on 25th January 1939 after the nearby River Gipping overflowed its banks.

project name 'Farmer Deck' the plough works fitted large ridging bodies and rollers on a frame at the front of Churchill tanks. The ridging bodies were to provide a safe path for the tank tracks by turning two furrows 12in deep by 24in wide to clear land mines, and a very heavy-duty Equitine cultivator was to be towed behind to clear mines between the tracks. A proving test was arranged for army officials on Nacton Heath but the device was damaged in unloading and failed to impress the onlookers. It was back to the drawing board and, following field tests with a flail attachment on a meadow at the nearby village of Tuddenham, it appears that Ransomes made about two hundred mine-clearing devices for Churchill tanks in the months leading up to the D-Day landings.

With many employees either serving in the forces or involved in armaments production there was a serious lack of the necessary labour to assemble the equally important farm implements. The shortage of labour was partly overcome by using groups of people in outlying villages who were unable to get to the factory due to their 'location, domestic duties or other responsibilities'. These village groups, consisting mainly of the ladies of the parish, were supplied with benches and hand tools so that they could work in their own homes and sheds. They undertook various tasks including salvaging, sorting and re-threading 70 tons of nuts and bolts, putting nuts and washers on thousands of harrow teeth and assembling plough wheels, handles and drawbars. The ladies at the village of Playford did simple fitting and de-burring work in their own homes

Men and women worked side by side in the Plough Works during the Second World War.

This Churchill tank was modified in the Plough Works for mine clearance work.

and the Waldringfield village group assembled drawbars for Motrac ploughs. Mr Pearce at Bredfield and other local blacksmiths undertook more skilled work including drilling and welding components for ploughs and other implements. The village groups completed over 80,000 hours of work and the women, in many cases no longer young, handled assemblies weighing up to 90lb.

The Orwell Works staff also played their part in the war effort in other ways. The company had its own fire brigade with over 200 officers and men, and the 11th battalion of the Suffolk Home Guard with 480 Officers,

The thrasher works in the early 1940s.

A village group assembling tractor plough drawbars during the Second World War.

NCOs and men was made up entirely of Ransomes' employees. The Home Guard Company mounted a nightly guard at the Orwell Works for over three years and was noted for its rifle-shooting prowess.

With the war at an end a 265-acre factory site on scrubland at Nacton on the outskirts of Ipswich was purchased in 1945 and the new iron foundry was completed in 1949. The press shop, plough assembly works and other departments were gradually transferred to the new site and on completion of the move in 1966 there were 3,200 employees at Nacton. The Orwell Works were sold in 1968. The new foundry had sufficient capacity to make anything from a tiny casting to the massive counterweight for Ransomes' electric trucks that took a week to cool down. The press shop had the largest drop hammers outside Sheffield. There was a huge gear-cutting department and another made wooden crates used to pack anything from a single-furrow plough to a combine harvester for the export market.

Ransomes and the Ford Motor Company signed an agreement in 1945 that led to the joint manufacture of a range of FR mounted tillage implements for the E27N Fordson Major at Ipswich and at the Ford factory at Leamington Spa. The arrangement came to an end on 1st January 1955 and from that time all FR implements and spare parts were made at Ipswich. The Ransomes name was used for the British market, and implements designed jointly with the Ford Motor Company also carried the FR brand name. The Ford Motor Company sold a range of Ipswich-built implements in some overseas countries under the FR Nacton brand name but Ransomes was the only acceptable name in some Middle Eastern countries. On the other hand, at one time Spanish farmers would only buy a Ransomes combine if it carried the Ford badge. Some export customers wanted a name rather than a model number on Ransomes implements, so to overcome the problem the TS 54 share plough, for example, was sold in some countries as the Ransomes Robin.

Ransomes' battery electric trucks were originally designed to carry loads or tow trailers but increasing demand for forklifts initiated the start of new era in the truck department in 1947 with the launch of a battery electric fork truck. Another era came to an end in 1948 when Ransomes made their last horse plough, and within a couple of years Punch and Rex, the last team of horses kept at the works for testing and sometimes demonstrating horse-drawn implements, were finally pensioned off.

Ransomes ploughs have been, and still are, used at countless numbers of ploughing matches in all many parts of the world. The YL body is a particular favourite for competition ploughing with horses and trailed or mounted tractor ploughs. The R.S.L.D. and similar trailed ploughs are also still popular and, where allowed, they trail all manner of ironmongery behind the mouldboards in a bid to win a first-prize card. Ransomes TS 86 match ploughs, with rather more adjustments than an everyday model, were a bit special. Any competitor using a TS 86 in the 1950s, 1960s and 1970s could expect to get more than a little attention when Ransomes ploughing match enthusiasts such as Jim Gass, Ernie Roworth and Paddy Burne were on the field.

Tractor-mounted FR reversible ploughs appeared in 1950 but in spite of a long experience with threshing machines the first Ransomes combine harvester, based on a Swedish design, did not appear until 1954. Low-density pick-up balers were added to Ransomes' product range in 1951. High-density wire-tying stationary balers based on a Ruston design had been made at Ipswich for several years before Ransomes acquired D. Lorant Ltd at Watford in 1951. Mr Lorant had been importing low-density balers from Claas in Germany since the late 1940s and after fitting wheels and tyres they were sold as Lorant balers. The business continued as Ransomes, Sims & Jefferies [Watford] Ltd until 1956.

The Steel Case Co. at Tredegar in Wales, which supplied Ransomes with steel plate for mouldboards and various case-hardened components, was acquired in 1953 and a marketing agreement for forklift trucks was signed with Hyster in America in the same year.

Thousands of prizes have been won with Ransomes horse and tractor ploughs at local, National and World ploughing matches.

Ransomes' catalogue for the 1954 Royal Show included ploughs, disc harrows, cultivators, toolbars, the MG motor cultivator, subsoilers, potato lifters, thrashers, balers, sprayers, crop driers and combine harvesters. In round figures, lawn mower sales for that year included 42,500 hand mowers, 9,500 motor mowers and 1,100 gang mowers of various types.

The 1960s saw the grass machinery division improve its existing machines and develop many new models, from small domestic machines to gang mowers for sports grounds and golf courses. A tractor-mounted, power-driven five-unit gang mower, launched in 1964, was another world first for Ransomes. Sales of grass machinery exceeded those of tillage equipment for the first time in 1973 and within six years accounted for 50 per cent of Ransomes' total sales revenue. The Danish-built

Ransomes Spares available here.

Nordsten seed drills were added to Ransomes' product range in 1967 and the Catchpole Engineering Co. Ltd at Stanton in Suffolk and Johnsons of March were acquired in 1968. The Ransomes range of root-crop machinery had been limited to potato ridgers, potato spinners and sugar-beet lifters until the purchase of these specialist root-crop machinery manufacturers. The Catchpole Engineering Co. were well known for sugar-beet harvesting machinery including the Cadet and self-propelled Powerbeet, while the Johnson range of potato machinery included planters and complete harvesters. Following the modernisation of Johnson's Engineering works, the welding of lawn mower cylinders and lawn mower bottom blades, previously bought in from Sheffield, was transferred to March in 1973. Component production and assembly of various implements including HR disc harrows, Tillorators and sprayers were moved to the March factory in 1974.

Ransomes, Sims & Jefferies identified 130 acres of land surplus to the requirements at Nacton in 1972 so the company established Ransomes Property Developments Ltd to develop a new industrial park. The plan was to build new factories and warehouses on the site during the following ten to fifteen years, and it was forecast that about 4,000 new jobs would be created in the Ipswich area. About 3,000 people were employed at the Nacton works in 1976 but within five years, when the annual turnover was in the region of £42m, the work force had been reduced to 2,400 employees. Combine manufacture was running down in the mid-1970s when the main products included ploughs, tillage equipment, root harvesters and sprayers. On the other hand, grass machinery had now become a major part of the business.

Wisconsin Marine was founded at Milwaukee in the early 1940s to manufacture marine salvage equipment but by the early 1970s they were only making Bob-Cat snow throwers. Fortunes changed with the launch of Bob-Cat rotary mowers in 1975; they were introduced to the UK market in 1977 and within a year Wisconsin Marine was a leading manufacturer of professional rotary mowers in America. Ransomes acquired a 30 per cent stake in the Milwaukee company in 1978 with the option to become the major shareholder by 1980, and in due course Wisconsin Marine became Ransomes Inc. with a factory at Johnson Creek in Wisconsin.

The Dorman Sprayer Co. at Ely was acquired in 1978 and, in view of the high costs involved in developing a new sugar-beet harvester to replace the ageing Catchpole machines, Ransomes became the UK importer for the Danish-built TIM sugar-beet harvesters in 1980.

Sales of grass machinery exceeded those of farm

machinery for the first time in the late 1970s and by 1980 only 40 per cent of the Nacton Works' £42m turnover was earned by the agricultural machinery division. Although their total workforce had fallen to 2,400, in 1980 Ransomes were still by far the largest agricultural and garden machinery manufacturers in the UK. Sperry New Holland at Aylesbury came second with 750 employees and a turnover of £17m. At least one-third of Ransomes annual output was exported in 1979 when, for example, disc ploughs were at work in Venezuela and disc harrows were preparing land in East Africa for maize. Motor Triple mowers were cutting parkland in Vienna and Mastiff motor mowers were used to trim lawns surrounding the Royal palace at Bangkok in Thailand. Ransomes Grass Machinery [Scotland] Ltd had a depot in Edinburgh and there were distribution companies in Australia, Germany and France. Other group interests included Ransomes Overseas Services selling technical knowledge, Ransomes Leasing Ltd and Ransomes Property Developments Ltd.

Eight years later in 1987, when Ransomes were the largest lawn mower manufacturer in Europe and third largest in the world, they sold the farm machinery division to the Electrolux Group who used the Agrolux name for their new business. Agrolux established their head office at Bluestem Road on Ransomes Park and the sale agreement allowed them to use the Ransomes name and designs for a limited period. The launch of a new range of Agrolux ploughs from Sweden in 1989 saw the end of Ransomes' plough production exactly two hundred years after Robert Ransome had started at St Margaret's Ditches in Ipswich.

Agrolux merged with Overum Tive in May 1991 to form Overum Ltd and the new company distributed a range of Swedish-built Overum farm machinery from their premises in Bluestem Road. Overum transferred their UK business to Newton Aycliffe in County Durham in January 1998 but within six months they in turn had merged with Kongskilde and moved back to the new Danish owner's premises at Holt in Norfolk.

Ransomes had returned to the domestic mower market in 1985 with the acquisition of G. D. Mountfield of Maidenhead, followed by Westwood Tractors of Plymouth in 1989. The Mountfield business was moved to Plymouth and merged with Westwood in the same year to form Ransomes Consumer Ltd.

Steiner Turf Equipment Inc. of Ohio and BTS Green with factories in France and Italy were bought in 1988. Further investment in 1989 added the North American turf equipment manufacturers Cushman and Ryan of Lincoln, Nebraska and Brouwer Ltd of Keswick, Ontario to the Ransomes group of companies. After two hundred and nine years of trading in Ipswich, Ransomes, Sims & Jefferies Ltd were acquired by the American Textron Corporation in 1998, and Ransomes Consumer Ltd became an independent company.

Danish-built TIM sugar beet harvesters were added to Ransomes' product range in 1980.

Chapter 2

Ploughs

Wooden ploughs with a mouldboard to turn the soil were used in Britain as early as the first century AD. By the early 1700s improvements included iron shares and coulters and the use of metal plates to cover the mouldboard. Iron-framed ploughs with wrought-iron mouldboards were gradually coming into use by the early 1800s.

Robert Ransome gained experience in plough making while serving his apprenticeship. He used this knowledge to make his own wooden-framed ploughs with iron soil-wearing parts when he started up in business at Norwich. Robert had already patented a process for tempering ploughshares with salt water when more by accident than design he discovered a way to make self-sharpening ploughshares. Some molten iron, accidentally spilled on the foundry floor, had cooled quite quickly where it came into direct contact with the cold floor, with the result that the underside of the iron was harder than the top surface.

When this principle was applied to the making of plough shares by using an iron chilling block in the sand mould, the softer top surface of the share wore more rapidly than the underside and so retained a sharp cutting edge. Robert Ransome was granted a patent for self-sharpening shares in 1803.

Almost every county had its own particular pattern of plough in the early 1800s and these designs remained virtually unchanged for centuries. Ransomes patented the AY wooden-framed plough based on a design used in Rutland in 1836 and by 1840 they were making more than eighty different types such as the Lincolnshire, Bedfordshire and Rackheath (Norfolk) each designed to suit the varied soil conditions in the British Isles. The Bedfordshire plough had wheels

The Rackheath Plough, invented by Sir Edward Stracey, Bart and made by Ransomes.

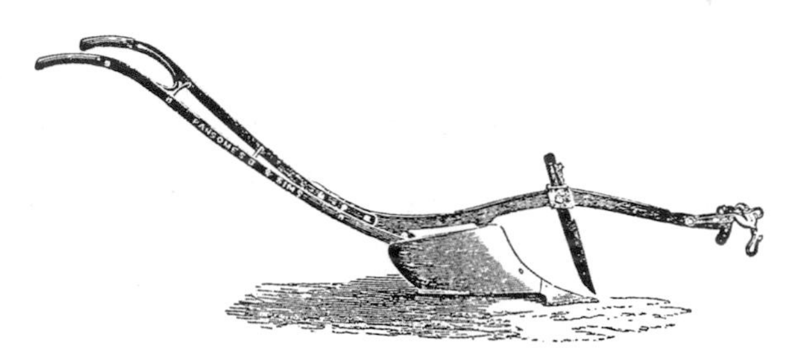

The S.C.W. solid-beam Scotch plough with a wrought-iron frame for use with up to six horses was included in the Ransomes & Sims 1862 catalogue. The S.C.W. cost £5 without wheels or £5 5s with one wheel. Cast-iron shares cost 10s a dozen and steel shares were 6s 6d each.

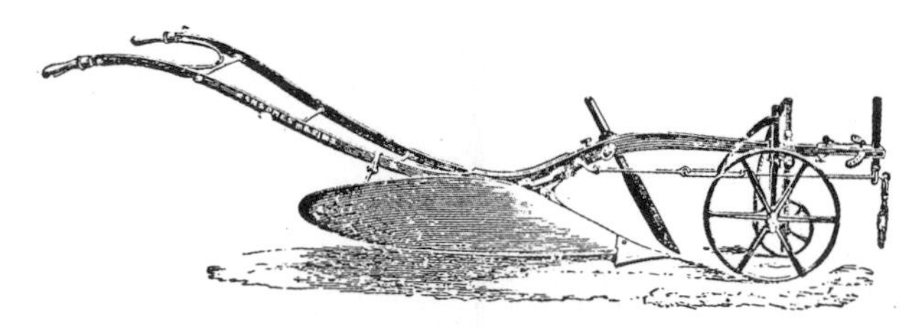

The V.R.S. improved solid-beam wrought-iron plough was 'strong enough for a team of eight or even ten horses or bullocks'.

with a lever for the ploughman to regulate the depth of the share while ploughing. The two-furrow Lincolnshire plough was said to work very effectively and 'in ordinary cases could produce a saving of a man and two horses for two acres of work'. The wheeled Rutland plough was of simple construction and light draught while the Rackheath plough was suitable for sub-soiling light land and 'stirring the soil under the sod in turf land'.

The change to iron-framed ploughs started in 1843 with the introduction of the improved Rutland plough with an iron beam and handles and the famous YL (Yorkshire Light) for general ploughing. The new YL plough body was used on Ransomes horse ploughs for the next hundred years and was still available on tractor ploughs in the 1980s. Steel-framed ploughs were in common use by the 1860s and like other British companies Ransomes were selling many different models to their farmer customers at home and abroad. The company had built a considerable export market for their products by the 1840s when horse, bullock, oxen and even elephant ploughs were in use in many parts of the world.

Ransomes & Sims' illustrated catalogue and price list for 1862 included a number of wooden- and steel-framed ploughs. The V.R.S. improved solid-beam, single-furrow plough with a wrought-iron frame cost £5 5s without wheels; one wheel added another 5s and two wheels were an extra 10s. The V.R.S. plough and the V.R.S.C. with a cast frame were supplied with or without wheels and there was a choice of a cast-iron or steel mouldboard with a wrought-iron or patent chilled cast-iron share. According to the catalogue these large ploughs 'will turn a furrow 12 in by 18 in' and they were 'strong enough for a team of eight to ten horses or bullocks'.

Ransomes were making rotary furrow-breaking attachments for horse ploughs in the late 1800s.

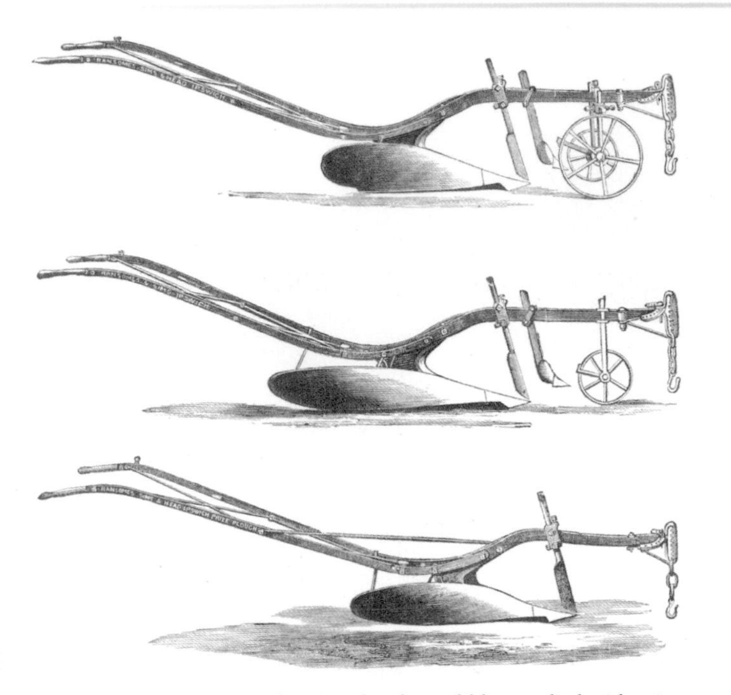

Ransomes' Newcastle prize ploughs could be supplied with a 'great variety of breasts, long, medium or short, made extra deep for crossing fallows and deep ploughing and suitable for light, medium or heavy soil.' A swing plough had a shorter beam and longer handles than a wheeled plough.

Two wheels were recommended on level land as they enabled a boy to do the work, because once set, the plough would run almost without holding it. A single land wheel was said to be better on unlevel or sticky soil but it would require great skill on the part of the ploughman and it would require his constant attention to guide the plough. The swing plough without wheels required even more skill as the ploughman had nothing to regulate the width or depth of the ploughing, although the long handles 'gave great command over the plough'.

Reversible ploughs are not a modern invention. Ransomes & Sims were making Joseph Skelton's patent S.P.T. iron turnwrest plough in 1866. The two mouldboards were used alternately; a lever was used to turn one of them into work while the other was raised and carried clear of the land. A second lever was used to move the knife coulter into the correct position and the wheels ran alternately in the furrow and on the land. Ploughs with alternative bodies for heavy and light land, designated the S.P.T.H. and S.P.T.L., were added in 1868. Light wooden-

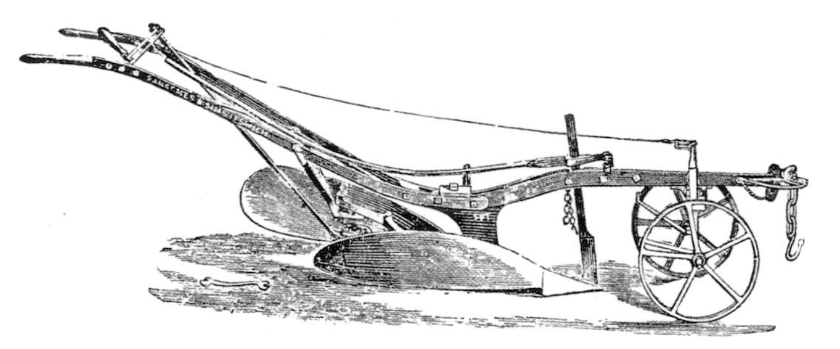

The S.P.T. iron turnwrest plough was recommended for ploughing up corners and other pieces of land inaccessible to a steam plough.

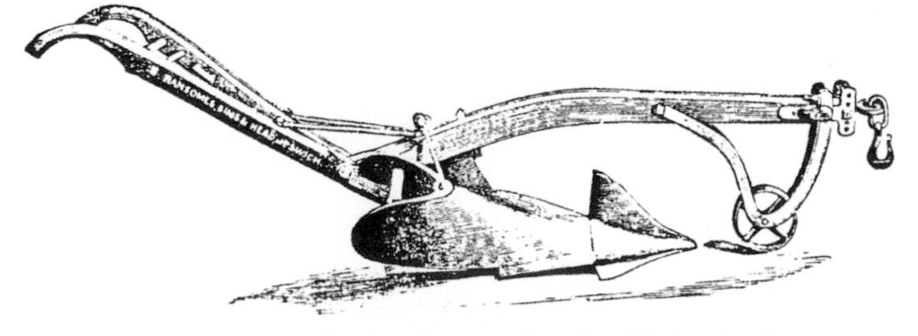

The mouldboard and share were turned under the wooden beam on this simple design of Ransomes turnwrest plough with the wing on the share serving as the coulter when ploughing in either direction.

Four of the six first prizes offered at a trial of iron-framed horse-drawn ploughs arranged by the Royal Agricultural Society of England at Newcastle in 1864 were won with Ransomes ploughs and for the next eighty years the Newcastle name was associated with horse ploughs made at Ipswich. About sixty different iron- and wooden-framed ploughs including lightweight pony ploughs, sulky or riding ploughs and ploughs for two to six horses or for two or three pairs of oxen were listed in Ransomes' 1886 catalogue. Five different size and weight Newcastle series ploughs were described in the catalogue as 'the most perfect which R. S. & J. have ever brought out'. Upward of one hundred different patterns of mouldboard and shares were available for these ploughs making them suitable for all types of soil, classes of work and local requirements. The catalogue explained the advantages of the three designs of Newcastle plough in some detail.

The Brabant turnwrest plough for two or three horses was described as 'strong and simple in construction, steady in work and light in draught'.

beamed turnwrest ploughs with a single mouldboard and the Brabant turnwrest plough were also made by Ransomes, Sims & Head in the late 1860s. The Brabant design, with one body in the air and the other in work, is still in use today.

Steam power was coming into use on some large farms by the end of the 1840s and in 1850 Ransomes made a mole plough to John Fowler's patent design. It was pulled back and forth across the field by a portable steam engine with a system of anchor pulleys and wire rope. In 1855 Ransomes' works manager William Worby met John Fowler while on holiday at Brighton. This chance meeting led to the manufacture in 1856 of a steam-powered four-furrow balance plough for Fowler at the Orwell Works. The plough was drawn by cable across the field using a Ransomes portable steam engine and anchor carriage. It was tested at the nearby village of Nacton and Fowler was so impressed by its work rate of one acre an hour that he ordered a Ransomes portable engine and plough on that very day. The plough was entered in the Royal Agricultural Society of England trials at Chester in 1858 and awarded the £500 first prize. The 10nhp portable Ransomes engine had a winding drum under the boiler and John Head designed the anchor carriage. The engine and anchor carriage wound themselves along the headland at the end of each pull. Ransomes & Sims made engines and ploughs for John Fowler until 1862 by which time he had established his own business at Leeds and introduced a more efficient ploughing system using two engines with cable drums to haul the balance plough back and forth across the field.

Steam power was fine for wealthy estate owners but most farmers still ploughed with horses. Some American farmers were already using riding or sulky ploughs and Ransomes' customers were able to follow suit in the mid-1880s with the single-furrow sulky plough with an automatic power-lift mechanism. The catalogue explained that 'the arrangement of the apparatus is such that the plough is raised entirely by the forward motion of the horses without any effort on the part of the ploughman.

Two or three horses were needed to pull the Ransomes sulky or riding plough with an automatic power lift mechanism. The sulky turned a furrow 5 to 8 inches deep and 14 to 16 inches wide.

The principle of the Jefferies lift was also used on Ransomes' M.E.D.M. two- and four-furrow chain-pull ploughs. The long handle was used to drop the plough into the working position (top) and set ploughing depth. With the wheels lowered, the plough was raised completely clear of the ground and could 'be turned on its wheels with perfect ease'.

'The driver was also able to vary the depth of the furrow while the plough was in work. Large numbers of Ransomes sulky ploughs were exported to Argentina.

Mouldboards were made from wrought iron until the late 1800s when steel ones came into use. Most horse ploughs used on British farms had a single mouldboard, as multiple-furrow models could be rather heavy for the ploughman to lift from work. The Jefferies Patent Double Wheel-Lifting Apparatus, awarded a silver medal at the 1885 Inventions Exhibition in London, overcame the problem. The Jefferies design consisted of a cranked axle with a pair of wheels mounted centrally on the frame. The

This two-furrow plough for three or four horses was raised from work with the Jefferies Patent Double Wheel-Lifting apparatus. A hand lever, connected to a pair of wheels on a cranked axle, was used to lift or lower the plough at the headland.

wheels on the cranked axle were lowered with a hand lever to take the plough out of work. When the wheels were raised to put the plough back into work, one of them controlled ploughing depth and the other was held clear of the ground.

Farm implement literature in the late 1800s provided far more information than can be found in their modern glossy counterparts. Multiple-furrow ploughs with the Jefferies mechanical lift were claimed to save time as there were fewer headland turns and a great deal more work could be done in a day. To back up this claim one Ransomes catalogue even gave details of how far a ploughman had to walk to complete an acre of land with different furrow-width ploughs. It took 12.33 miles and 6 hours 10 minutes to complete an acre when ploughing 8in wide furrows. This was reduced to 9.9 miles and 4 hours 57 minutes with a 10in furrow, and of course with a multiple-furrow plough there was a drastic reduction in the distance walked to plough an acre.

Fewer models of horse ploughs were made in decreasing numbers at the Orwell Works in the late 1930s. Newcastle ploughs, one- and two-furrow light R.H.A. ploughs for smallholders, together with the I.R.D.C.P., T.C.P. and G.D.N. digger ploughs were included in the Ransomes 1938 implement catalogue. Ploughs specially designed for Scottish conditions and already made for many years were also listed. They included the Claymore digger and the Guidtop, so-called because it set up the furrows with a good [guid] top. Cast-iron shares were quite unsuitable for many Scottish soils with heavy stones but Ransomes' bar point pattern ploughs were satisfactory. The Gudekut with a steel bar point and steel wing was specially designed for the Highlands and the Gudekut S with a steel bar point and cast-iron wing was made for the Lowlands. The Galloway, introduced for use on stony soils in southern Scotland, had the same mouldboard as the I.R.D.C.P. share plough but with a steel bar point. There was also a bar point option for the Newcastle series ploughs.

Tractor Ploughs

Although production of animal-draught ploughs for home and overseas markets continued to thrive, an increasing number of oil-engined tractors on farms in the early 1900s prompted Ransomes to introduce a deep-digging chain-pull trailed tractor plough in 1904. It was similar to the American design of riding plough with a seat for a second man to operate the controls and to lift and lower the plough into and out

Riding ploughs were the next stage in the development of the modern tractor plough but two men were needed to plough a field. This R.Y.L.T. plough, made in 1914, is hitched to a Ransomes 4nhp compound steam tractor built in 1909.

Tractor ploughing became a one-man job with the introduction of the R.S.L.D. with a self-lift mechanism operated from the tractor seat.

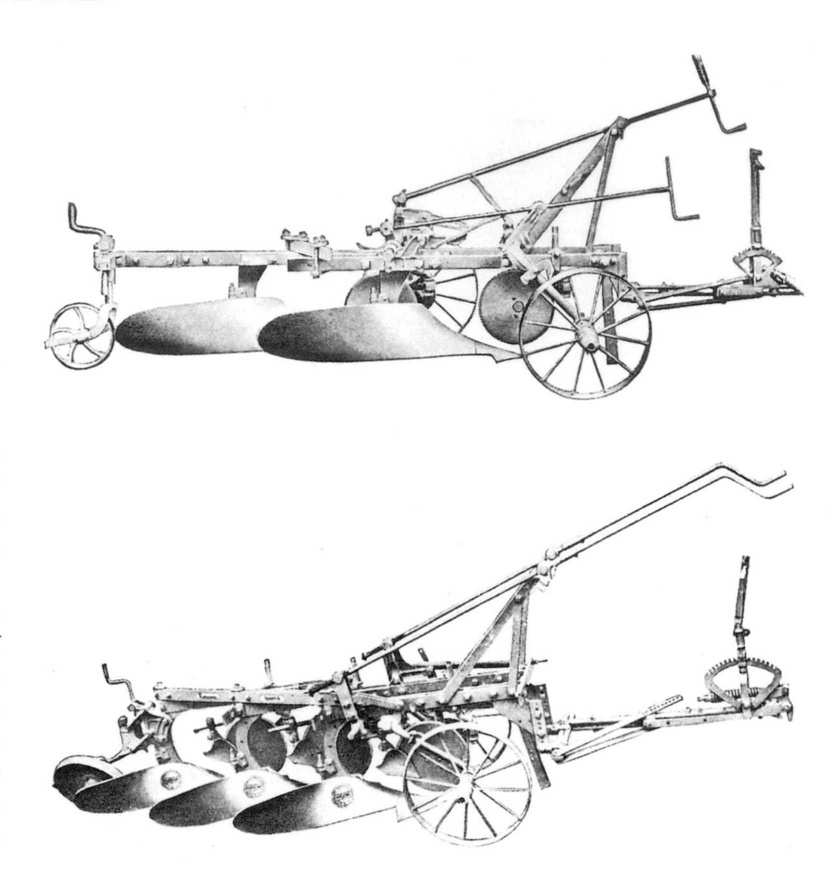

The Motrac, introduced in 1927, had a clutch-type self-lift mechanism.

of work. The steerable two-furrow Yorkshire Light Tractor Multiple [Y.L.T.M.] appeared in 1909, the Ransomes Motor Tractor Double and Motor Tractor Multiple [R.M.T.D. and R.M.T.M.] ploughs were added in 1913 and the three- and four-furrow R.Y.L.T. in 1914. They too were similar to Ransomes' horse-drawn sulky models, as the tractor ran on the land and a seat was provided on the plough for the steersman.

Much of the Orwell Works was turned over to the production of munitions at this time to help with the war effort but Ransomes also made tractor ploughs with YL or digger bodies for the War Department. Many of them were used with the MOM tractors, which were really Fordson Model Fs, imported from America by the Ministry of Munitions. Ransomes celebrated the end of the war with the launch of the Victory animal-draught plough and many thousands of them were exported for use on sugar plantations.

Tractor ploughing became a one-man operation with the introduction of the R.S.L.D. No.1 [Ransomes Self-Lift Double] and R.S.L.M. No.1 [Ransomes Self-Lift Multiple] in 1919. The main controls were within reach of the tractor driver and a rope-operated rack and pinion mechanism raised and lowered the plough into and out of work. The ploughman tugged on the rope to engage the rack with the pinion on the land wheel to lift the plough frame and bodies from work, and when it was fully lifted the frame was locked in the raised position. A second tug on the rope lifted the rack from

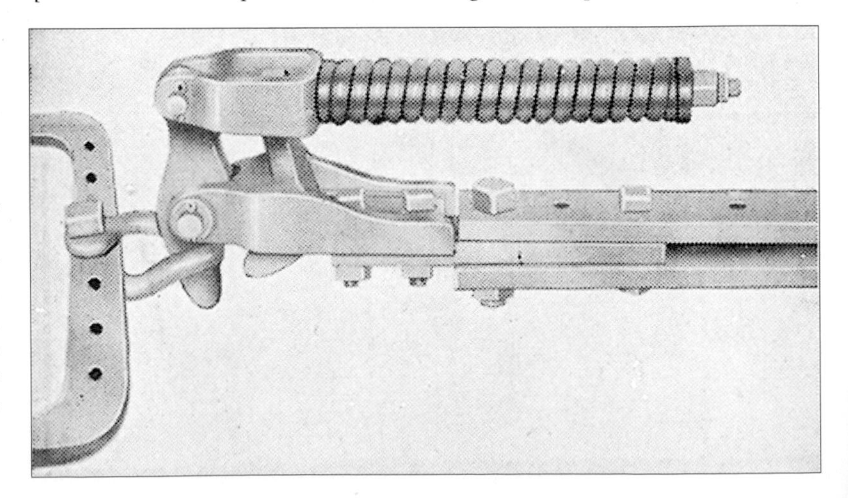

Many Ransomes ploughs, including the Motrac and R.S.L.D., had a spring-loaded safety drawbar hitch that automatically released the plough if it was seriously overloaded or hit a tree root or other obstruction.

The Weetrac was the first Ransomes mounted plough.

the pinion and the bodies dropped back into work.

Several versions of the R.S.L.D. and R.S.L.M. were made before they were eventually discontinued in the early 1950s. The No.1 had hand levers to adjust furrow width and depth but the R.S.L.D. / R.S.L.M. No.2 and later models were adjusted by screw handles. The No.4 had a stronger frame with optional bar point bodies. The No.7 and the heavier No.8, both with two or three furrows, were listed in the early thirties. The two- and three-furrow No. 9 and the four-furrow No.11 for ploughing light land had replaced them by the end of the decade. The No.12 appeared in 1946 and the R.S.L.D. / R.S.L.M. No.15 with the choice of seven body types was the last of the line.

A host of new models appeared in the late 1920s. The Weetrac, introduced in 1927 with general-purpose bodies, was Ransomes' first mounted plough designed for the increasing number of tractors with a mechanical lift linkage. The two- and three-furrow Motrac with forged beams and legs also appeared in 1927 and the Midtrac, a heavy-duty version of the Motrac, originally designed for Scottish farmers, was added in 1929. The

The six-furrow TS39A Hexatrac for crawler tractors weighed about 1½ tons.

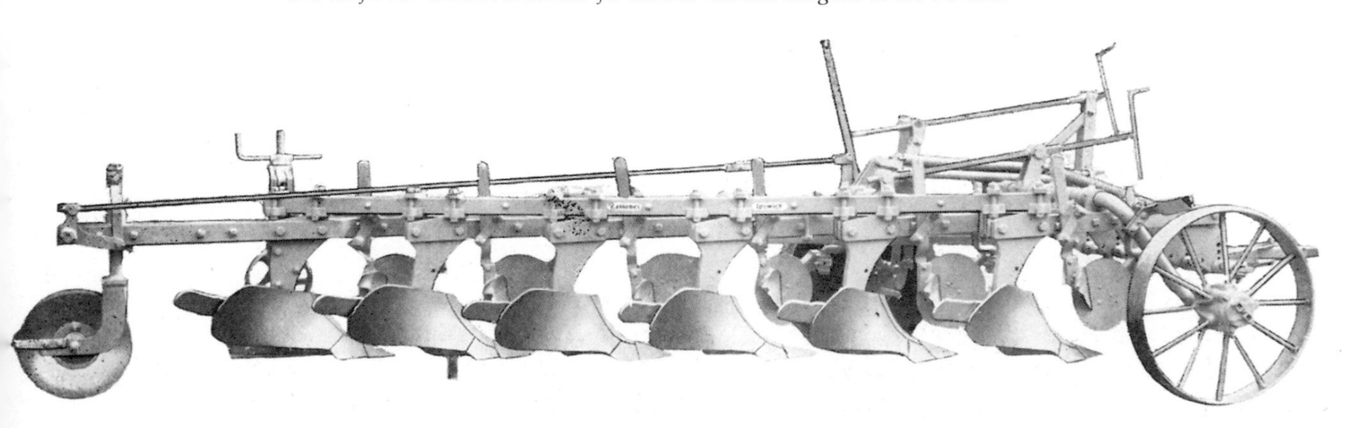

three- and four-furrow Multitrac, derived from the Motrac and designed for crawler tractors, appeared in the same year. Increased tractor power brought the launch of the five-furrow Quintrac in 1931 and the six-furrow Hexatrac completed the family in 1936.

Introduced in 1927, Ransomes Riffler ploughs with six shallow working bodies were designed for 'paring fields after harvest, for cross-ploughing, cleaning fallows and ploughing for barley on thin soil up to 4 inches deep.' Equipped with sword shares, the Riffler turned 8in wide furrows but one body could be removed and the remaining five were equally spaced to give a 9½in furrow width. It had a furrow width adjuster on the drawbar and two screw handles to adjust the depth and keep the plough level.

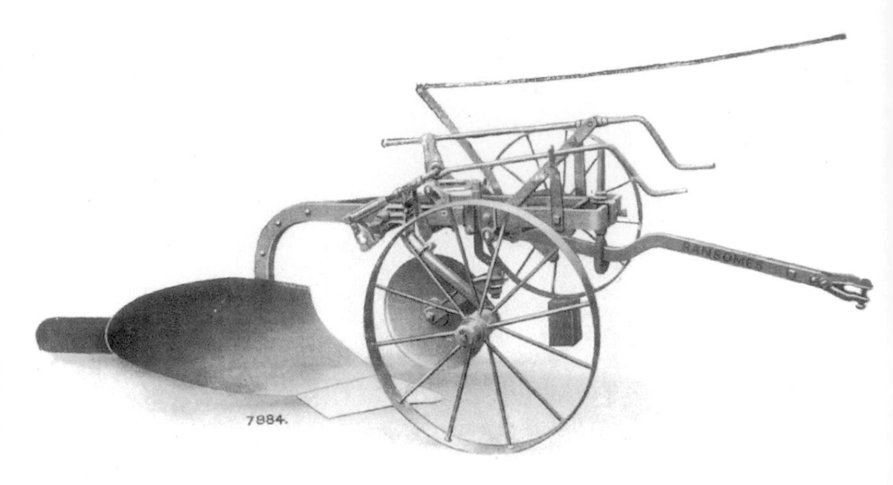

Unitrac deep digger ploughs, which cost £26 10s in 1932, could be set to turn furrows between 14 and 18 inches wide and up to 16 inches deep.

Riffling is an East Anglian term for breaking stubbles after harvest and the Ransomes Riffler plough was made for this work.

Balance Ploughs

Ransomes had added horse-drawn balance ploughs to their range of turnwrest ploughs by the early 1930s. There were various types for one or two horses, some of which had a steering lever, while others were steered with a pair of handles. The horse was either unhooked at the end of each run before the opposite body was lowered or, with some difficulty, it could be manoeuvred round to the opposite end without unhitching the plough.

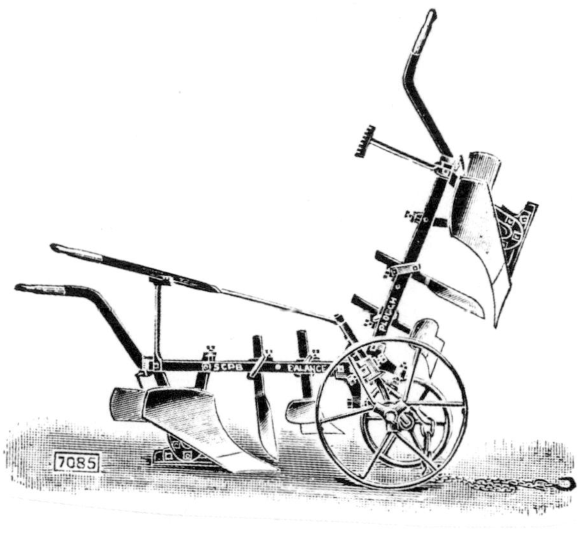

Ransomes' S.C.P.B. deep-digging horse-drawn balance plough was steered with one handle bar and the lever connected to the land wheel. The lever was repositioned after the opposite body was lowered into work ready for the next run.

In 1932 Ransomes acquired the agricultural machinery business of James & Frederick Howard Ltd of Bedford who had been one of Ransomes' strongest competitors. The Howard catalogue for that year pointed out that their implement range was now manufactured by Ransomes, Sims & Jefferies Ltd. The products included various ploughs and cultivators, disc harrows, horse hoes and hay rakes. The Howard balance plough cost £14 11s 6d complete with knife coulters, and an optional pair of skim coulters were an extra 21s 6d. An alternative balance plough with 'Kent breasts for ploughing up to 9 inches deep' cost £18 17s 6d. Suitable for ploughing between 4 and 7in deep, 'it turns the furrows all one way', the catalogue explained, 'and lays them in the most approved form.'

Five Twinwaytrac two-furrow balance or one-way ploughs were made in 1935 and 1936. They were similar in design to the earlier steam balance ploughs with one pair of bodies in work and the other in the air. Pulled by a chain from a crawler tractor drawbar, the Twinwaytrac worked to a depth of 16in. The tow chain was dragged under the wheel when the tractor turned on the headland and the heavy central

The Twinwaytrac, single- and two-furrow, one-way ploughs were described as a one-man outfit for use with track-laying tractors. The land wheel had a rim extension and an over-tyre was used when travelling on the road.

Steel wheels were often used when tractor ploughing in the late 1940s.

castings, made by Lake & Elliot, were designed to overbalance the airborne bodies and drop them into work when the tractor set off on the next run.

The TS 25 two-furrow trailed plough with a working depth of 6in and designed for the new 6hp Motor Garden [MG] crawler was also launched in 1935. It was not too difficult to convert the hand-lift TS 25 down to a single-furrow plough and increase furrow depth to a maximum of 10in. The two-furrow plough cost £12 and the single-furrow model was £9 9s. The self-lift TS 30 and TS 31 single- and two-furrow ploughs for the MG 2 appeared a year or so later.

Ransomes had manufactured at least three hundred different types of share and disc ploughs to meet the needs of their customers at home and abroad by the early 1930s when they were still producing many types of mouldboard to satisfy their world-wide market. The AD was a deep-digging body for the Argentine; special bodies were produced for the French market and various mouldboards for animal-draught ploughs were made for sale in distant lands.

Ransomes were still making horse ploughs and about twenty different types of tractor plough in the late 1930s but at the start of World War II in 1939, and in spite of strenuous opposition from many of their farmer customers, the range was considerably reduced for the duration of hostilities. Plough production was given top priority in the push to grow more food and to this end Ransomes were making No.3 Motracs at a rate of 130 a week in 1943. Animal-draught ploughs, including the single-furrow Victory, were made at Ipswich throughout the war and production for the home market eventually came to an end in 1948.

The drive to convert swords back to ploughshares at the end of the war resulted in the Ford Motor Co. and Ransomes signing an agreement for the joint manufacture and sale of FR and Fordson implements for the E27N Fordson Major. Ford had bought the Flavel gas appliance iron foundry at Leamingtom Spa in 1940 with plans to produce their own range of implements but the war intervened and the factory was used to produce tracks for military tanks.

The first Ford Elite trailed ploughs, with Ransomes bodies, were made at Leamington in 1945. The Ford Motor Co. also made mounted ploughs with cast steel frames at Leamington with Ransomes bodies from the Orwell Works. In 1946, in order to compete with the Ford plough, Ransomes introduced the two- and three-furrow FR PM [Plough Mounted] plough that had a heavy cast-iron headstock and a choice of YL, EPIC and EFR bodies. The single-furrow TS 55, launched in 1950, gave farmers the choice of a single-, double- or three-furrow Ransomes plough for the E27N Major. The TS 55 had more under-frame clearance than the PM series and a lighter new fabricated headstock.

The Ford Motor Co. made the two-furrow EP2J and three-furrow EP3G mounted ploughs with a cast headstock and Ransomes bodies at Leamington for the 1951 launch of the New Fordson Major. Unfortunately the three-point linkage geometry was not quite right, resulting in problems with fractured headstocks which was solved by replacing them with a stronger fabricated headstock.

This 1951 advertisement for the TS 58 Midtrac Major illustrates the high standard of ploughing used for Ransomes' publicity material.

Although Ransomes were very much involved with the Ford Motor Co., they also made mounted ploughs for other tractors. In 1948 they announced the TS 54 [Robin], a mounted plough without a depth wheel suitable for the Ferguson TE 20 and the American Ford 8N, and in direct competition with the two-furrow Ferguson mounted plough normally bought by farmers ordering their first Ferguson tractor. Ferguson's general-purpose mouldboards were not popular in some areas and local Ferguson dealers modified the frogs and sold these ploughs with Ransomes YL mouldboards and shares. The TS 54A with a depth wheel was made for category I linkage tractors without depth control and the TS 54E, also with a depth wheel, was for tractors with category II linkage. The two- and three-furrow TS 54B and TS 54D with depth wheels were later versions of the plough.

With the exception of the No.15 R.S.L.D. and R.S.L.M. all Ransomes' trailed tractor share ploughs were given a TS model number in the late 1940s. The TS 43A Motrac, TS 45 Midtrac, TS 46A Multitrac and TS 39A Hexatrac were among the nine trailed TS ploughs listed in the 1950 edition of the 'Hitch a Ransomes to your Tractor' catalogue. The TS 27A Unitrac Minor was a digger model with a UD body designed to take a furrow 12in wide and up to 12in deep. The single-furrow TS 16E Unitrac Major was a heavier version for crawlers and large wheeled tractors. It could turn a furrow 14in to

The FR PM two- and three-furrow ploughs were the first of a long line of mounted ploughs made at Ipswich for Fordson and later for Ford tractors.

A depth wheel was not required on the TS 54 plough for the Ferguson TE 20 and Ford 8N tractors.

18in wide and up to 16in deep! British farmers were buying mounted ploughs in the mid-1950s but Ransomes were still producing ten trailed models, mainly for export. The TS 44 Litrac, TS 6 Duotrac, TS 45 Midtrac, TS 46 Multitrac and the TS 3 Proconsul were for general work. The TS 47 Duratrac launched in 1950, the TS 1 Solotrac, TS 4 Giantrac and TS 41 Supertrac, designed for the Indian market, were deep diggers turning furrows up to 16in wide and up to 18in deep.

Reversible Tractor Ploughs

Renewed interest in reversible ploughing prompted the launch in 1950 of the mounted single-furrow TS 51 and the two-furrow TS 50 which were awarded an R.A.S.E. Silver Medal at the 1952 Royal Show. The reversing mechanism on both ploughs was operated with a long hand lever attached to the tractor's rear axle. The same manual turnover mechanism was used on the two-furrow TS 68 and single-furrow TS 74 which superseded the TS 50 and TS 51 in the mid-fifties.

Disc ploughs, mainly for export, were still being made at the Orwell Works in the 1950s and three new models appeared in 1954. The two- and three-furrow TD 17 adjustable for 8, 9, 10 and 12in furrows cost £95 and £115 respectively and the three-furrow TD 16 mounted reversible disc plough was £130.

Ransomes ploughs were regular winners at the annual World Ploughing Championships and they triumphed yet again in 1954 when the first, second and third prize-winners all used ploughs made at Ipswich. The 1950s were busy times in the Ransomes plough works, as Fordson tractors were sold in huge numbers and conventional right-handed FR ploughs were still in demand. The first TS 59 ploughs were made at Nacton in 1953 and within a year the

The FR TS 1016 for the Fordson Dexta was the first Ransomes reversible to have a hydraulic turnover mechanism.

Ransomes mounted plough range included the two- and three-furrow TS 59 [Raven], the TS 55 [Falcon] single-furrow deep digger for reclamation work and the TS 50 and TS 51 reversibles. The TS 64 [Vulture] two-furrow with semi-digger or digger bodies and the two-furrow TS 63 described as 'ideal for match ploughing' completed the range. Draft control hydraulics were not yet available on Fordsons but they could be supplied with a pre-set linkage control and Ransomes introduced the four-furrow TS 73 without a depth wheel in 1955.

The FR TS 1013 two- and three-furrow and TS 1014 single-furrow right-handed ploughs for tractors with category I linkage appeared in 1955 and 1956 respectively and were made for five years. The launch of the Fordson Dexta tractor with category I hydraulic linkage and 'Qualitrol' draft control at the 1957 Smithfield Show coincided with the introduction of the single- and two-furrow TS 1015 and 1016 reversible ploughs. A double-acting ram (DAR) and hydraulic control valve on the Dexta were used to turn the TS 1015 and TS 1016 reversible ploughs to the left or right. By the autumn of 1958 Ransomes were building an average of fifty mounted implements every day of the week to meet the demand for ploughs and tillage equipment for the Fordson Dexta.

The two-furrow TS 82 reversible with a self-loading mechanical turnover unit operated with a hand lever, launched in 1961, and the three-furrow TS 83, introduced in 1962, were both made until 1970. The new Ford 1000 series arrived on the scene in 1964 and the heavy-duty three-furrow TS 84 was announced at the same time. Within a couple of years the mechanically reversed 80 series reversible plough range had grown to include the single- and two-furrow TS 80 and TS 81 for 30 to 55hp tractors with category I or II linkage. The TS 82, TS 83 and TS 84 were designed for more powerful tractors with category II hydraulics. The TS 81, TS 82 and TS 83 were discontinued in 1970 but the TS84 remained in production for another four years.

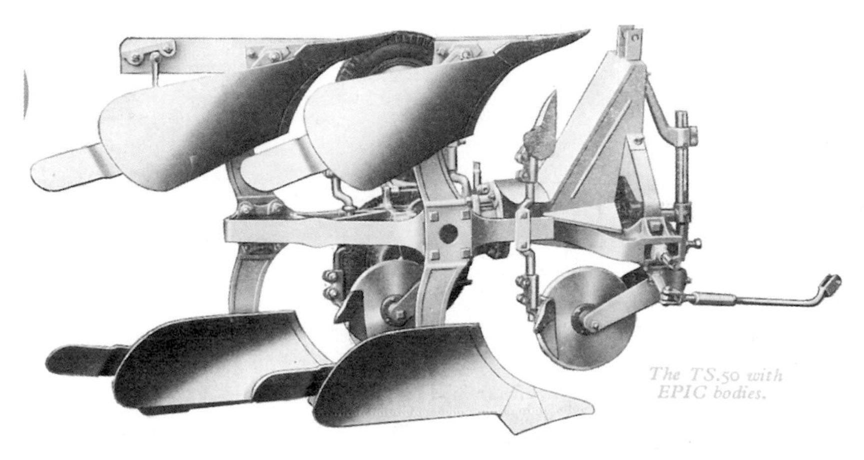

A long hand lever was used to reverse the bodies on the FR TS 50 reversible plough. Publicity material explained that the plough was so well balanced that it turned virtually under its own weight and little or no manual effort was required to reverse the bodies.

Hitching the FR TS 51 single-furrow plough to an E27N Fordson Major was a one-man operation.

Wooden grinding cradles, used at the turn of the century to hand-grind and polish mouldboards and other soil-engaging parts, were still in use in the early 1950s.

The first automatic mouldboard grinders and polishers were installed in 1952.

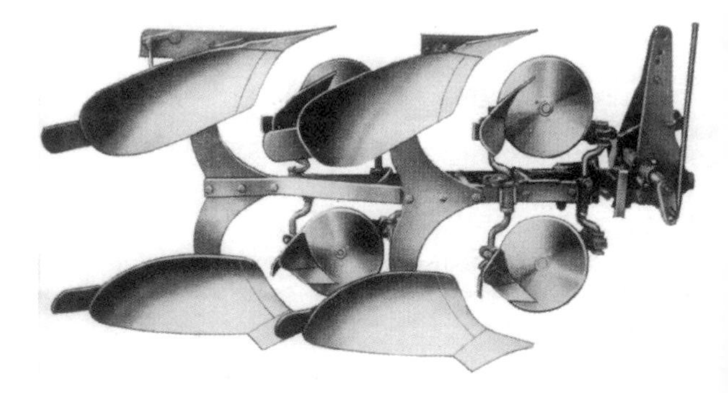

The TS 82, introduced in 1961, was the first of a popular range of mechanically reversed Ransomes ploughs.

The New Theme ploughs were introduced to the farming public at the 1967 Royal Smithfield Show.

Reversible plough sales had outstripped the demand for conventional models by the mid-1960s when a reduced range of Ransomes right-handed ploughs included the two- to five-furrow TS 59, TS 64 and semi-mounted TS 78. Tractor power was increasing at the time but as Europe's plough makers had failed to produce reversible ploughs with enough furrows to use the extra power, there was a brief return to mounted multi-furrow right-handed ploughs.

The four- and five-furrow TS 88 mounted plough specially designed for the Marshall crawler tractor was launched in 1967 and the New Theme two- to five-furrow TS 90 and TS 91 right-handed ploughs superseded the TS 59 and TS 64 ranges in the same year. The largest five-furrow plough provided a more realistic load for the County, Roadless and other four-wheel-drive tractors of the day. The New Theme ploughs had hollow box-section steel beams; TCN high-speed bodies taking a furrow 10in wide and up to 10in deep were standard, with YL bodies available on special order. Farmers who bought one of the smaller FR TS 90 ploughs could use Ransomes' beam exchange scheme to add an extra furrow or two for their more powerful tractors by trading in the existing beam for a longer one. The two-, three- and four-furrow TS 91 turned furrows 12in wide and up to 12in deep. A 14in TS 91 was added in 1968 followed by the four- and five-furrow TS 95 and the four-, five- and six-furrow TS 94 in 1969. The six-furrow version, with UCN bodies capable of ploughing a 7ft width of land up to 10in deep in a single pass, provided a realistic load for the new breed of 100hp four-wheel-drive tractors which were popular in the early 1970s.

Ransomes had a new competitor in the reversible plough market in the early 1970s when Warwickshire farmer Roger Dowdeswell designed and built a reversible plough with Ransomes bodies in his workshop. Neighbouring farmers liked the plough and asked him to make one for them. Dowdeswell ploughs gained in popularity and although many

The FR TS 96, launched in 1971, was a five-, six- or seven-furrow semi-mounted plough with YL or UCN bodies for high horse-power tractors with wheels running on unploughed land to prevent soil compaction in the furrow bottom. The rear of the plough was raised and lowered on the headland with a remote-controlled hydraulic ram.

A considerable number of the two hundred or so plough body types made by Ransomes during their 200-year history were designed to meet the needs of a particular export market or individual customer. However, by the late 1930s they were making fewer than twenty different bodies.

YL *Yorkshire Light for whole furrow work up to 9in deep. First made in 1843, it was superseded by the YCN in 1985.*

EPIC *Semi-digger body with furrows up to 10in deep for more broken finish.*

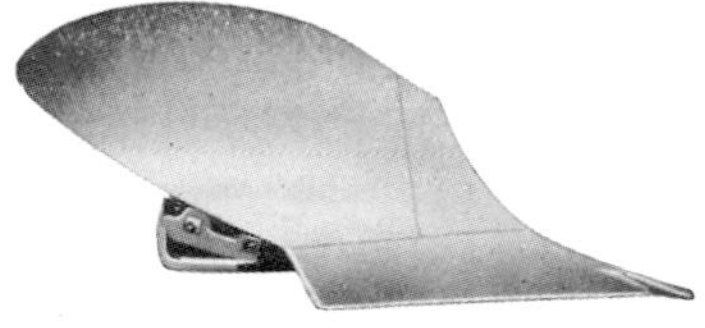

DMC *Semi-digger body up to 9in deep and 10, 12 or 14in wide furrows.*

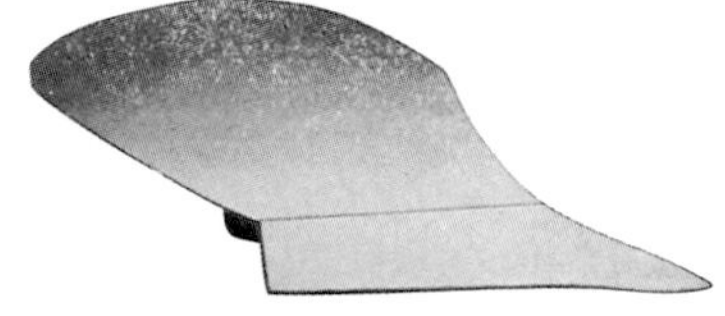

AD *Deep digger designed for the Argentine.*

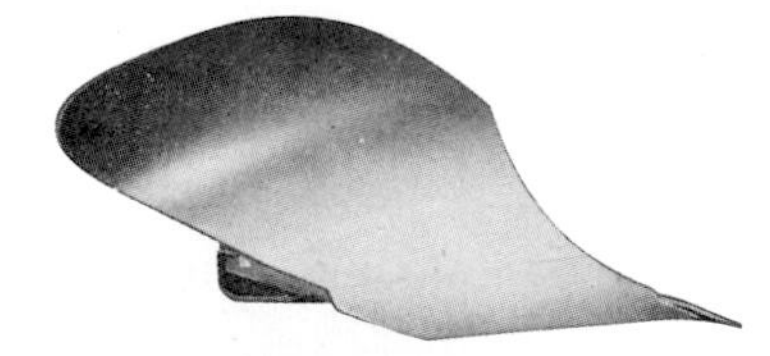

UD *Large semi-digger body for work up to 12in deep and 12in wide.*

UN *Export digger body 16in wide and 10 to 8in deep.*

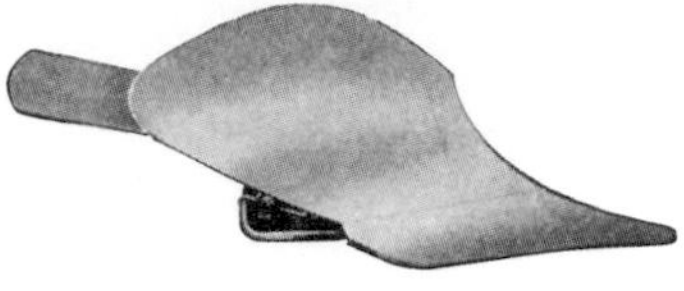

UDM *Large digger body ploughing up to 16in deep.*

TCN *Multi-purpose body working up to 10in deep and 12in wide.*

UCN *Turns furrows 6-10in deep and 12 or 14in wide. Turns unbroken furrows at low speed and semi- or fully broken work at higher speeds.*

SCN *Semi-digger / digger ploughing 8 to 12in deep.*

SL *Swamp body for boggy land.*

UG-BP *Bar-point deep digger for furrows 16in wide and 18in deep. Worked best when ploughing deeper than the furrow width.*

farmers bought one of the new green ploughs, Ransomes did at least make money from the sale of plough bodies and spares to their competitor. Ransomes also sold plough bodies to Ernest Doe & Son who used them on ploughs they built for the Triple-D four-wheel-drive tractor.

The three-furrow TSR 108 was added to the 100 series reversible ploughs at the 1974 Smithfield Show. With 40in between the bodies and a 26in under-beam clearance it was ideal for ploughing in straw and was advertised as 'the plough that breaks through where others break down'. Disc coulters, considered extra equipment for many ploughs by the mid-seventies, were not an option for the TSR 108 but it did have skimmers and a new type of sword landside already popular on the continent. The TSR 112 introduced in 1976 with auto-reset trip legs with hydraulic cylinders connected to a gas/oil accumulator for rocky soils was one of several variants of the TSR series.

The Spaceframe 200 conventional mounted ploughs, made between 1975 and 1985, was the last series of Ransomes, Sims & Jefferies right-handed or fixed ploughs. The designers used the concept of building bricks where a relatively small number of components could be used to make up a plough to suit virtually every farmer's needs. There were short and long versions of the eight basic models with two to six furrows, high-clearance beams, two inter-body clearance options and the choice of shear bolt, mechanical trip leg or hydraulic reset trip leg overload protection. The shorter model with 30in inter-body clearance was suitable for general ploughing while the long version with 40in between the bodies was recommended for ploughing in trashy conditions. A semi-mounted conversion kit with a hydraulically operated rear wheel unit was available for the larger ploughs. UCN bodies were standard, furrow width could be varied with a turnbuckle arrangement on the frame and there was an optional hydraulic front furrow width adjuster.

Eight versions of the TSR 100 series mounted reversible plough for tractors in the 50 to 140hp bracket were in production by 1981. A 75hp tractor was recommended for the two-furrow TSR 102 with manual turnover and disc coulters and for the three-furrow TSR 103 with optional hydraulic reversing mechanism. The three-furrow TSR 104 and TSR 105 reversibles appeared in 1978. The lightweight TSR 104 was suitable for tractors with only 65hp under the bonnet and sales literature suggested the TSR 105 with a side-shift facility built into the headstock to cater for varying tractor wheel widths required a 75 to 95hp tractor. The two-furrow TSR 107, three-furrow TSR 108 and three- or four-furrow TSR 108A ploughs designed for difficult working conditions had the largest inter-body clearances of the TSR range. Sword landsides were a feature of the five-furrow TSR 109A launched in 1978, which had a depth wheel behind the third furrow and transport wheel at the rear.

The semi-mounted five-furrow TSR 150, introduced in 1980, was designed for tractors of up to 180hp. Bolt-on beam extensions for a sixth and seventh furrow were available; the legs were protected with shear bolts, and there was a choice of disc coulters or sword shares with a vertical cutter.

The Experimental Department

The experimental department was definitely out of bounds to the casual visitor, and to safeguard my informants they can only be identified as Geoff, Mark and Rick. The department evolved from a blacksmith's shop where the smith's skill was used to make and assemble new implements. Designers with their drawing boards and instruments came next. Prototypes were made, tested in the field and modified several times before working drawings were eventually passed to the shop floor and put into production. Then, from the 1960s, computer-aided design made drawing boards and tee squares just as obsolete as the blacksmith's forge. However, it always came down to testing a new machine in a variety of field conditions to decide whether it would make it to dealers' showrooms.

For many years plough design was based on the principle of 'make it and then test it to see if it was possible to break it', and if it survived the test it was probably too strong. In later days technology took charge. Strain gauges were used to calculate loadings on implements both in work and in transport and an Austin Gypsy was fitted out with recording equipment to print out the data on 'toilet rolls'. The drawing office sometimes designed plough beams using steel that was too good for the purpose. It was sometimes strong enough but it was too flexible and prototype ploughs suffered from sagging beams, resulting in uneven furrow depth. At other times the beams might deflect sideways to give unequal furrow widths.

Stress testing did not always find the weak points and a special circular track complete with bumps and ruts was built at Nacton and used to test mounted implements non-stop day and night, if need be to destruction. A four-wheel drive Roadless tractor attached to a control arm from the centre of the test track provided the power to test, for example, the wear rate of the head stock and turnover mechanism of a reversible plough.

There was great secrecy when taking out a prototype for a field test. The first Potentate, a self-propelled potato harvester which failed to reach the production stage, was driven from the works covered with camouflage netting. Unfortunately it became wedged on a narrow bridge and next morning the Development Shop charge hand on his way to work was greeted by a neighbouring farm hand who had seen Ransomes' new 'spud harvester' at the farm.

Repetitive testing could be a chore at times and one bright lawn mower tester saved himself a lot of shoe leather by tying a motor mower to a tree with a long length of rope which took about forty-five minutes to wind round the tree. Then, after turning the mower round, it took another 45 minutes to unwind the rope again with no need to walk behind. At other times, even in the early 1980s, the driver was a necessary part of the test programme. A number of disc coulters, running on ball bearings, were mounted on a toolbar and the tractor was driven up and down on sandy heath land at Nacton. This simple trial proved that the tried and tested oilite bushes had a far superior working life to ball bearings.

There were times when the experimental staff had a limited time to develop and deliver a new machine. A government contract to build six ground nut harvesters to lift the crop and pick the shells from the roots was completed in six weeks. A team, drawn from the plough works and combine department, worked seven days a week with two twelve-hour hour shifts each day to finish the job on time. The harvesters were delivered, the crop was lifted, picked and bagged up with reasonable success. Unfortunately no one had allowed for the high moisture content of ground nuts and before long the sacks overheated and then caught fire.

Field staff were very much involved in testing prototypes on farms and the farmers concerned quite rightly looked upon them as experts in their field. On one occasion a farmer who had a problem with his plough, wrote to the company asking them to 'send that man with a hammer because he knows where to hit the plough to make it right again.'

There were times when the Experimental Department probably wished the camera had not been invented. A design fault resulted in a fractured hydraulic pipe and the plough ended up in a heap on the ground

Push-pull ploughs were fashionable in the early 1980s. Between six and nine furrows could be ploughed in one pass with a two- or three-furrow front-mounted TSR 300 FD and a standard TSR 300 plough on the rear linkage.

Although the TS 109A was designed as a five-furrow model, bolt-on beam extensions meant it could be converted to a three- or four-furrow plough. Introduced in 1977, the TSR 112 was designed for ploughing the rock-strewn soils found in Scotland and the north of England. A single wheel served as a combined depth and transport wheel, automatic reset legs were standard equipment and the hydraulic accumulator built into the system had sufficient capacity to re-set two bodies at the same time. Designed for high-speed ploughing, the three-furrow TSR 113, four-furrow TSR 114 and three-furrow TSR 111 convertible to four furrows were launched in 1980; optional equipment included sword shares and spring-loaded auto-reset trip legs for ploughing land where large stones could be a problem.

The TSR 300 series superseded the TSR 100 series in 1982. The standard TSR 300 reversibles had two, three or four bodies and the heavy-duty TSR 300 HS was a four- or five-furrow plough. They had UCN or SCN bodies and disc coulters were standard on the rear bodies. The three- and four-furrow TSR 300 D had a longer frame to accommodate disc coulters and skimmers on all bodies. The growing interest in push-pull ploughs prompted Ransomes to introduce the front-mounted two- and three-furrow TSR 300 FD in 1982. The TSR 300 LD with greater inter-body clearance and the 300 LT with trip legs were added in 1984. Optional trash boards, which could be bolted to the mouldboards to improve burial of combine straw, were available for these ploughs.

The five- and six-furrow TSR 300 HD with a new headstock design to prevent the rear wheel hitting the ground when reversing the bodies and reduce the load on the hydraulic linkage was launched at the 1985 Smithfield Show. Other new products at the show included the YCN body, a modern version of the famous YL mouldboard, suitable for ploughing 4 to 8in deep, a range of slatted mouldboards and a furrow press for the TSR 300 series ploughs.

Plough mouldboard design at Ransomes had always been a matter of trial and error. The head blacksmith shaped a mouldboard to meet a particular purpose that was repeatedly tested and re-shaped until the required result was achieved. The YCN body was, according to publicity material, probably the first body designed in Britain with the aid of computer technology.

Ransomes Agrolux

After Ransomes' agricultural machinery division was sold to Electrolux in 1987, the new owners carried on with the production of the existing range of ploughs. Then Agrolux launched a new range of ploughs made by Overum in Sweden at the 1989 Smithfield Show. Although Ransomes painted some of their export machines a bright orange, most of their ploughs and other implements were blue, originally the light shade of Orwell blue. The darker Nacton blue was used from the late 1940s but the new Agrolux range was painted bright red to distinguish it from the Ransomes ploughs that had been made at Ipswich for the past two hundred years.

The new Agrolux ploughs included two-, four- and five-furrow reversibles, and two- to eight-furrow, mounted and semi-mounted conventional right-handed ploughs with auto-reset or shearbolt-protected legs. In 1991 Agrolux merged with Overum, who marketed a range of Swedish-built two- to eight-furrow Overum Ransomes ploughs in the more familiar blue livery. Overum moved to County Durham in 1998 and within a few months they joined forces with Kongskilde and the business was transferred to Holt in Norfolk.

Agrolux introduced the Red Plough range in 1989. The red paintwork was the first new colour in forty years apart from the change from Orwell blue to Nacton blue.

Chapter 3

Tillage Implements

Although the Ransomes name is synonymous with ploughs, the company was equally well known for their tillage implements. Biddell's wooden-framed scarifier, invented by a Suffolk farmer of that name, was exhibited on Ransome's stand at the very first Royal Agricultural Show of England held at Oxford in 1839. The English Agricultural Society's *Journal* described the scarifier as a machine with 'two rows of tines fixed to a strong iron frame carried on two average-size wheels and preceded by a smaller pair.' The Journal also explained that 'chisel points are affixed to the tines which are removable and hoe blades of 4½ inches wide for partial hoeing and 9 inches wide to cut the land close may be substituted as occasions require.' The show judges praised Biddell's scarifier, noting that 'it had been constructed with much mechanised skill and power' and 'was able to break and stir eight acres in a day.' Ransomes made various sizes of Arthur Biddell's scarifier and otherhorse-drawn cultivators for the rest of the nineteenth century and later.

Ransomes' 1886 catalogue included a Scotch grubber, horse hoes, chain harrows and spike-tooth harrows. The handles and frame of the five-tined Scotch grubber were made from wrought iron. The chain harrows were between 5ft and 9ft 6in wide and in most cases 7ft 6in long. Light seeds, medium and heavy spike-tooth harrows were listed in the catalogue. One type had the teeth bolted to a zigzag frame while others had flexible joints that allowed the harrow frame to follow ground contours.

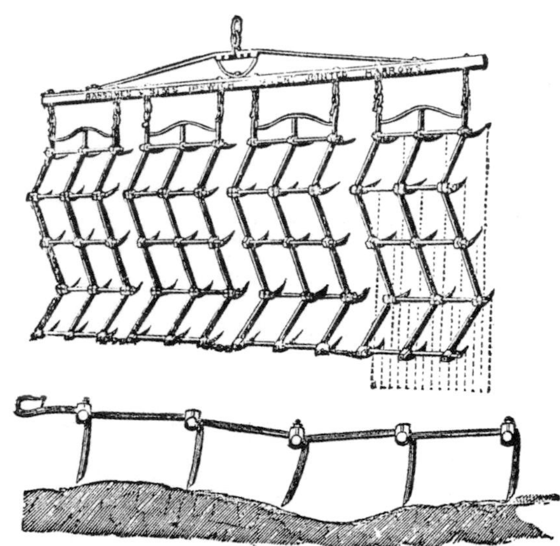

Ransomes' patent jointed harrows were made in two weights with two, three or four harrows in a set.

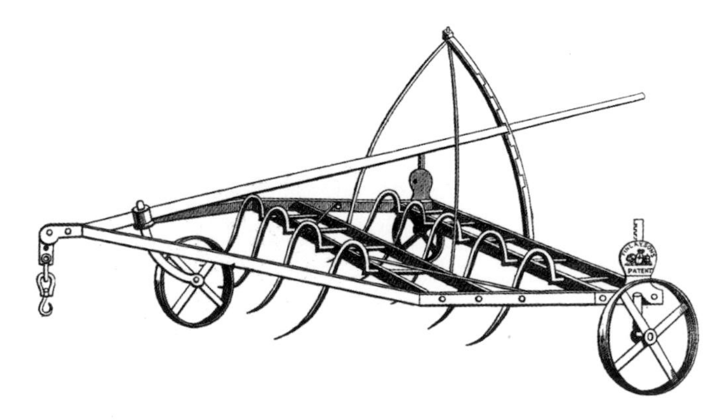

J.R. & A. Ransome were appointed makers and agents for Finlayson's patent self-cleaning harrows in 1832.

The 1886 catalogue recommended the use of the improved Scotch grubber for cleaning land. The grubber cost £8 15s and replacement points were 5s 6d a dozen.

Cultivators

A range of steel-framed horse-drawn cultivators with a cast-iron seat for the horseman was being made at the Orwell Works in the late 1800s. The two-horse Ransomes Small Holdings cultivator had a hand lever by the seat to put the seven rigid or spring tines into and out of work. The Orwell and Orwell Junior cultivators introduced in the early 1900s were made for about thirty years. The horse- or tractor-drawn Orwell could have seven, nine, eleven or thirteen tines with working widths from 5ft 3in to 8ft 2in. There were four hand levers on the Orwell cultivator, two of which were connected to the wheels mounted on cranked axles to adjust working depth. The third lever could be operated from the seat to raise or lower the tines while the final handle was behind the seat to be used when the horseman chose to walk behind the implement. Expanding axles were also available to accommodate the use of hoe blades or ridging bodies on the nine- and eleven-tine cultivators.

The Orwell Junior was similar to the Small Holdings cultivator with a single control lever. A head wheel, which took the weight from the horses' necks, was a standard fitting on the Junior and Small Holdings cultivators. A short length of chain with a hook, attached to the head or swivel wheel bracket at the front of the cultivator, was used to hitch the implement to a horse whippletree or tractor drawbar.

Ransomes' implement catalogue for 1932 explained that the Ipswich Steel cultivator was unequalled for breaking up turnip land after sheep and preparing seedbeds after ploughing. A single lever adjusted the working depth of the ten, thirteen or seventeen spring tines that were independently mounted to enable them to follow the contours of the ground. The Triplex cultivator, introduced in the early 1900s, could also be used as a horse hoe or ridger. A horse pole and whippletrees were recommended when hoeing root crops, and provision was made for the farmer to control the horses either while walking behind or riding on the implement.

Dauntless tractor cultivators were made between the late 1920s and the early 1950s. Ten models of Dauntless self-lift cultivator were included in Ransomes' farm implement catalogue for 1932. The No.3 and No.9 Dauntless cultivators had spring-mounted rigid tines but the No.1W, similar to the No.3, and the No.9W with smaller wheels, had fixed rigid tines. Special turf-cutting blades for aerating grassland were available for these cultivators. The heavier Dauntless No.11 with twelve rigid tines and the No.10 with fourteen tines were made for the larger models of wheeled tractor and for crawlers.

The even heavier Dauntless No.13A and No.14 could be converted to three-row ridgers or, with the rigid tines set in groups of three, could be used as scufflers.

The C5 and C13 with rigid tines were the heaviest models of Ransomes' trailed cultivators made in the early thirties. The nine-tine C5 with a 6ft working width weighed 12cwt and the 9ft C13 with fourteen tines weighed in at 17 cwt.

Small Holdings Cultivator fitted with spring tines.

The No.3 Triplex was a combined eleven-tine cultivator, ridger and horse hoe. It was controlled either from the seat or by a man walking behind. When used for hoeing or ridging, the Triplex could be set to work three rows spaced between 24 and 30 inches apart.

The nine-tine C28 and heavy-duty C17 cultivators together with the Dauntless No.15, No.16, No.19 and No.20 cultivators appeared in the mid-1930s. The No.15 and No.19 were lightweight versions of the No.3, and the No.16, designed for heavy cultivating, had thirteen spring-loaded tines and extra wide wheel rims.

Hydraulic linkage was standard equipment on most new tractors sold from the late 1940s and Ransomes made mounted toolbars for these tractors. However, there continued to be a good demand at home and abroad for Dauntless cultivators and fifteen different models from the No.1 to the No.24 were still being made in the early 1950s.

Mechanical implement-lifting devices for tractor-mounted toolbars were in use by the late 1930s and Ransomes followed this trend with their own design of mounted row-crop toolbar. This had a pair of cast-iron or pneumatic-tyred depth wheels and a ratchet lift mechanism that required just two or three movements of the hand lever to lift the toolbar. The frame was lowered by pressing the same lever forward to release a brake in the lifting gear, which allowed it to drop at the required speed. The toolbar and lift mechanism for the Farmall F12 and F20 cost £24 with cast-iron depth wheels; nine cultivator tines or three ridging bodies added another £8 10s and nine hoe blades cost £5 5s. A mechanical power lift, shaft-driven from the transmission, was optional equipment on the Farmall F20 and some other tractors. Farmers owning one of these tractors could buy a Ransomes toolbar without a mechanical lift unit for only £19.

Farm implement production at the Orwell Works was restricted during World War II and the continuing shortage of steel was reflected in the reduced range of cultivators made in Ipswich during the immediate post-war years. The

The Dauntless No.3 tractor cultivator had a rack and pinion self-lift mechanism on both wheels, and either wheel would lift it from work. The seven-tine cultivator had a 4ft 9in working width, while the nine- and eleven-tine models cultivated a strip of land 6ft 3in wide.

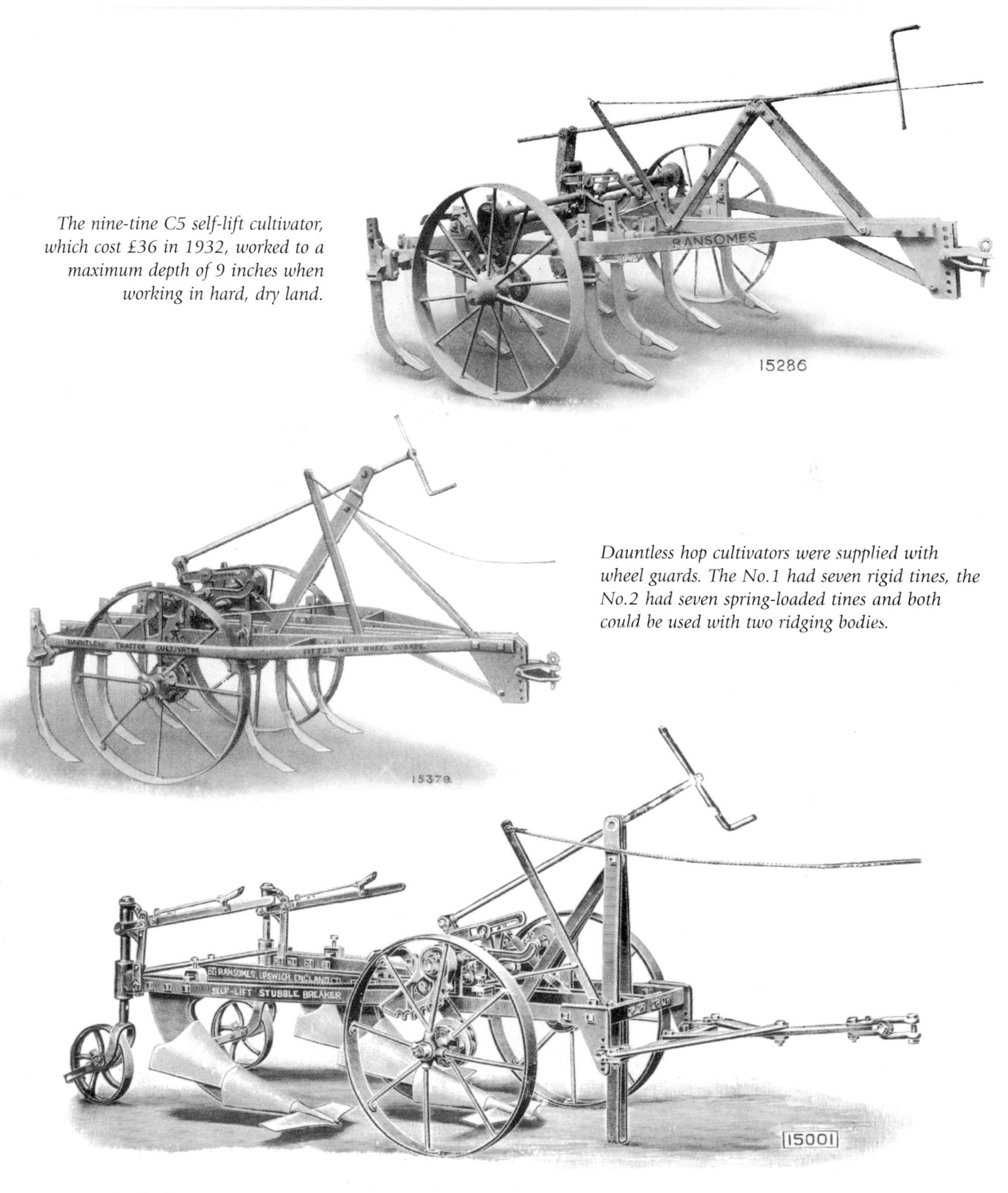

The nine-tine C5 self-lift cultivator, which cost £36 in 1932, worked to a maximum depth of 9 inches when working in hard, dry land.

Dauntless hop cultivators were supplied with wheel guards. The No.1 had seven rigid tines, the No.2 had seven spring-loaded tines and both could be used with two ridging bodies.

Ransomes' self-lift stubble breaker was used to 'break stubbles up to 4 inches deep and put the ground into small drills'. Ransomes' 1932 implement catalogue explained that 'this renders the weeds and rubbish subject to atmospheric conditions and ensures their extermination.'

relatively new tractor hydraulic systems brought a sharp decline in sales of trailed cultivators in the early 1950s. However, Ransomes were still making various trailed models including the Dauntless No.3, No.11, the new 10ft wide Dauntless No.16 with thirteen spring-mounted tines together with variants of the C13, C17, C18, C31 and C65. The C13 and C17 were still available in the mid-1960s.

The C13C heavy-duty cultivator for crawler tractors had a front or head wheel to support the drawbar. It had a rack and pinion self-lift on both wheels and the thirteen tines gave an 8ft working width. The letters A, B, C and so on after model numbers indicated a change in build, which may have been no more than an improved casting or forging. These letters were also used to denote a variation in the standard specification of an implement.

The tines on C18C and C31B Equitine cultivators, introduced in the mid-1930s, were designed to swing on their mountings. Each row of tines was connected by a system of chains running through pulleys attached to the tine bars. The tines worked the soil in the same way as an ordinary cultivator but when one of a linked set met an obstruction, the chain allowed it to swing backward and clear the blockage. The other tines in the group were pulled forward by the blocked tine and with the obstruction cleared the chain returned the row of tines to their working position. Later models of the Equitine had groups of three swinging tines linked by chains to compensating arms on the tine bars.

The C60 rear toolbar frame for the E27N Fordson Major was much more than a mounted cultivator. Farmers prepared to spend time bolting different attachments to the toolbar could use the C60 with rigid

Ransomes' mounted toolbars were made for the Farmall F12, F20 and F30, Case, Fordson, John Deere and Allis-Chalmers tractors.

The C17D with five rigid tines and the nine-tined C17E both weighed over a ton. Ransomes' catalogue for 1939 explained that they were designed for the most powerful tractors yet made.

or spring mounted cultivator tines, ridging bodies, sugar-beet lifting shares, potato-raising bodies or a steerage hoe conversion kit. The C61 front toolbar, also for the E27N Major, could be used with potato-covering bodies or a set of hoe blades.

The C60 and C61 toolbars were superseded by the rear-mounted C62 toolbar with an improved three point linkage headstock when the New Fordson Major was launched in 1951. Other rear-mounted Ransomes toolbars at this time included the C66 toolbar for Ferguson and Ford 8N tractors with depth control hydraulics.

Cultivator tines, ridging bodies, potato coverers and hoe blades were made for the C66, and with an optional steerage attachment and seat it became a rear-mounted steerage hoe. The heavy-duty C68 toolbar with tines and other attachments mounted on a 2¼ or 2½in square steel bar was also made for the New Fordson Major and similar tractors. The mounted C65A and C65B cultivators with seven and nine tines respectively, built

The C61 front tool frame was used for cultivating, ridging and hoeing row crops.

for the export market and designed to break up unploughed, sun-baked land up to 9in deep, completed the range of toolbars made at the Orwell Works in the early 1950s.

The C72 and C73 rear-mounted toolbars, launched at the 1955 Smithfield Show, were made for nine years. The 7ft wide C72 was designed for Ferguson and other tractors with category I hydraulic depth control systems, and the C72A for other category I linkage tractors had adjustable depth wheels. The 8ft 4in wide C73 toolbar was suitable for category I or II linkages and the 10ft 2in wide C74 was made for use with category II hydraulics. Attachments for the three toolbars included cultivator tines, ridging bodies, potato coverers, root lifters and disc bedders. The TCR 1001 and TCR 1004 toolbars with cultivator tines and other attachments for the New Fordson Major, Dexta and Super Dexta and the spring-loaded rigid-tine TCR 1003 Tiller were made for varying periods between 1956 and 1967.

Mounted spring-tined cultivators were either being made or imported by at least twenty different British companies when Ransomes introduced the Tillorator in 1959. The 7ft 6in TCR 1025 Tillorator with twenty-three spring tines was made for the Fordson Dexta and the 9ft 10in wide TCR 3006 with twenty-nine tines was matched to the Fordson Power Major.

Rigid and spring-loaded cultivator tines and ridging bodies were made for the FR C79 8ft 10in and 10ft 6in wide mounted 'Zed' section toolbars. Prices for the C79, launched in 1964, started at £84 10s. The new FR C81/96, C81/136 and C81/160 Tillorators with 4in tine spacings replaced the TCR 1025 and TCR 3006 in 1965. The second part of the model number used for the C81 series Tillorators indicated working widths of 9ft 6in, 13ft 6in and 16 ft. The wing sections on the C81/136 were folded by hand for transport and a hand-operated winch was provided for folding the wings on the C81/160. The 21ft 4in wide C81/214, added to the Tillorator range in 1968, had hydraulically folded wing sections.

The FR C83/126 ripper cultivator with square hollow-section beams was designed for 50 to 70hp tractors.

The triple-beam 8ft 6in wide C83/86 and twin-beam 12ft 6in C83/126 heavy cultivators were launched in 1966. Tines for heavy cultivating, stubble breaking, chisel ploughing or pan breaking were bolted to the hollow square-section beams, and the C83/86 could be used for subsoiling after removing the rear beams and bolting a pair of subsoiler legs on the front beam.

Designed for crawlers up to 80hp and four-wheel-drive tractors in the 85 to 120hp bracket, the 15ft 6in wide C87 mounted cultivator with a 4in square hollow-section frame and fifteen tines appeared in 1970. The C87 was a strongly built implement for stubble cleaning with heavy-duty rigid or spring tines, chisel ploughing and pan busting. A new design of heavy-duty tine called the Terratine was introduced in 1972 as an additional option for the C87 and earlier C83 series of cultivators. Terratines were said to have the combined advantages of straight and sloping tines to give maximum penetration with minimum draught. The leg was secured with a shear pin to prevent damage and the hardened steel points could be moved forward to compensate for wear.

The heavyweight C90 toolbar with category II and III linkages for four-

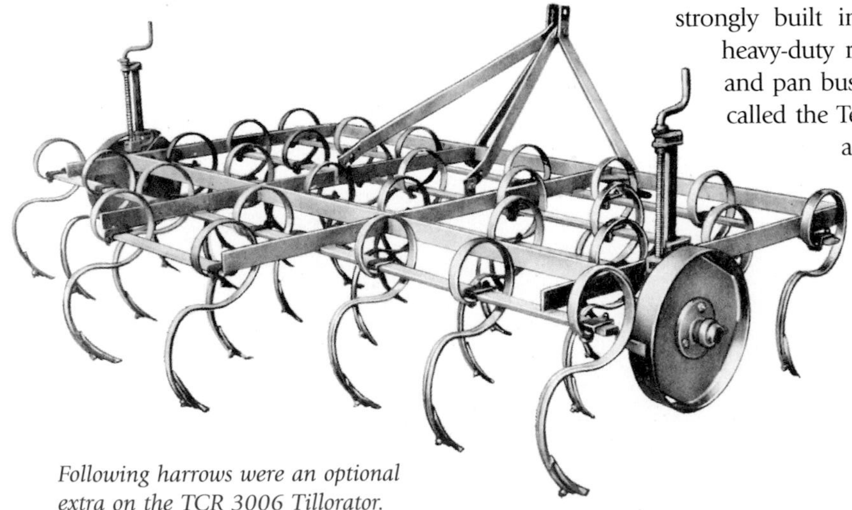

Following harrows were an optional extra on the TCR 3006 Tillorator.

wheel-drive tractors up to 135hp and 90hp crawlers appeared in 1971. Suitable for heavy cultivating, subsoiling and mole draining, the C90 had a massive 4½ x 7½in hollow box-section steel frame.

The C79 and C90 cultivators remained in production well into the 1980s but the C83 and C87 series were superseded in the late 1970s by the three-bar C92 and two-bar C93 cultivators. Designed for tractors in the 60 to 140hp bracket, the C92 series was advertised as 'a range of cultivators made to measure for your tractor and your land'. The C92 was put together in various sizes by bolting different wing extensions to the main frame. The smallest 8ft version had seven tines and the widest had twenty-three tines across a 17ft 4in toolbar. An automatic three-point linkage coupler and an end-tow transport kit were offered as optional extras on the wider versions of the C92 models. Various sizes of the two-bar C93 for tractors up to 100hp were assembled in a similar way.

The C95 Triple Task for 100 to 200hp tractors with category II or III linkage, introduced in 1980, was a heavy-duty, three-in-one cultivator with a box-section frame and optional crumbler roll. Made in 6ft 7in, 9ft 2in and 11ft 10in widths, the Triple Task had three rows of tillage tools working progressively deeper from front to back.

Disc Harrows

By 1890 horse-drawn disc harrows used, especially on ploughed-up grassland, to break up the furrows and produce a seedbed were being made at the Orwell Works. There was a seat for the horseman, and in order to produce a level finish the two sets of concave discs faced in opposite directions on their axles. By 1920 Ransomes were also making tandem disc harrows for tractor draught with four sets or gangs of discs mounted on individual shafts.

Ransomes No.2 King disc harrows were made for export. The 6ft wide harrow required six to eight oxen and it took a team of twelve to sixteen oxen to pull the 12ft machine.

Various types of spring and rigid tines for shallow and deep cultivations were used with the C92 cultivator.

An early 1930s catalogue suggested that the 8ft and 10ft wide tractor-drawn Baron tandem disc harrows with 20in diameter discs were ideal for preparing seedbeds on ploughed land. The same catalogue illustrated four sizes of Baronet tandem disc harrows with 22in diameter discs and working widths of 5ft to 9ft 6in. The discs were 9in apart and Ransomes recommended the Baronet for use on very dirty land and for turning in green manure. The 8ft wide Baron cost £54 in 1932 and the 9ft 6in Baronet was £57. Baron and Baronet disc harrows with the letters HR followed by a model number were made for the next twenty-five years or so. A catalogue for 1939 included the Baron No.3 [HR11 to HR15] and Baronet No.2 and No.3 [HR5 to HR9] tandem disc harrows and a two-gang Baronet with 22in diameter scalloped discs. Three models of Baron No.3 and Baronet No.3 disc harrows remained in production until 1967.

Production of the Howard No.2 Reliable tandem disc harrow as well as the tandem and two-gang lightweight Countess harrows was transferred to Ipswich in 1932 when Ransomes acquired James & Frederick Howard Ltd of Bedford. The acquisition enabled Ransomes to widen their disc harrow range as Reliable and Countess were both lighter in weight than the Baron and the Baronet harrows. The cutting angle of the front and rear gangs of the 6, 7, 8 and 10ft wide No.2 Reliable disc harrow was adjusted with a screw handle and an optional offset hitch was available for orchard work. The No.2 was superseded by the No.3 Reliable disc harrow in the late 1940s. The Countess harrow, suitable for tractor or animal draught, was mainly sold on the export market. (Page 168.)

Ransomes described the new mounted HR20 A-type offset disc harrow, introduced in 1950, as a 'modern harrow for a modern farmer'. Designed for the Fordson Major and similar tractors, the 6ft 6in wide HR20 had eighteen 20in diameter discs and weighed 8cwt. Its 'oil-soaked hardwood bearings were easily and cheaply replaced when worn'. An improved version of the HR20 designated as the HR20A (category II) and HR20B (category I) and made between 1955 and 1958, was equipped with a central stabilising disc to counteract the tendency for the two gangs to crab sideways. The FR THR 1005 mounted, A-type disc harrow was added to the range in 1956. It was specifically designed for the Fordson Dexta's draft-control hydraulic system, and an even penetration of both gangs of discs was achieved with an adjustable tie rod linking the rear gang to the headstock.

Towing trailed disc harrows on the road was always a problem until the introduction of the trailed 6ft and 7ft FR THR 1006 / 1007 in 1956 and the 8ft FR THR 3001 in 1958. These were equipped with hydraulically

The No.3 Baron and No.3 Baronet tandem disc harrows.

The HR20 A-type disc harrow could be offset by moving the headstock to either side of the frame.

raised and lowered transport and depth wheels with pneumatic tyres, making it possible at last to move disc harrows around the farm at speed. The smaller models were designed for the Dexta and the 8ft harrow was matched to the Fordson Major. The working depth was set with an adjustable stop on the double-acting ram and the disc gangs ran in iron bearings. The trailed THR 1006 and 1007 were discontinued in 1964 and the THR 3001 disc harrow was made until 1968.

The trailed, A-type Duchess HR21, 22 and 23 offset disc harrows with 7ft 6in to 10ft 6in working widths were introduced in the late 1950s mainly for the export market. They were made until 1966 when the HR33 and the heavy-duty offset HR34 A-type trailed

Spanners were needed to adjust the angle of the discs on the THR 3001 tandem disc harrow. A 60 to 70 horse power tractor was recommended for the 8ft HR 33 discs. Hydraulically operated transport wheels were standard equipment.

With an optional articulating top link arrangement, the HR44 disc harrows could be trailed while in work and lifted on the hydraulic linkage for transport and turning on the headland.

harrows designed for four-wheel-drive and crawler tractors replaced the Duchess 20 series models. The 8ft and 11ft wide HR33 with hydraulically operated transport and depth wheels was of similar design to the earlier THR 3001. F.W. Pettit at Moulton near Spalding made the offset HR34 series disc harrows. There were four working widths from 7ft 6in to 12ft and they could be end-towed on pneumatic-tyred wheels.

The HR29, 30 and 31 series of mounted disc harrows superseded the earlier 20 series harrows in 1962. The smallest 5ft 3in wide A-type HR29 cost £98 and the widest 8ft 6in HR31 tandem harrow was £132. The 7ft and 8ft 6in HR35 A-type mounted disc harrows together with the 10ft 6in and 12 ft HR36 tandem mounted harrows for 65 to 100 hp tractors with category II linkage appeared in 1969. These mounted and trailed, tandem and offset models with working widths of 7 to 12ft were made until the new 40 series for 60 to 100hp tractors was launched in 1978. The A-type HR42 and tandem HR44 were the first of the new 40 series mounted disc harrows. The HR42 had eighteen, twenty or twenty-two discs on a frame that could be converted with relative ease to a tandem harrow. The wider HR44 had twenty-eight, thirty-two or forty discs and the wing sections on the 12ft model were manually folded for transport. The HR46 A-type trailed harrow for 75 to 200hp tractors was launched at the 1979 Smithfield Show. The choice of twenty-six, thirty-four and forty-two discs allowed the HR46 to cover 10ft, 12ft 6in or 15ft 6in of ground in a single pass. The wing sections on the 34- and 42-disc models were folded hydraulically. Options for all 40 series harrows included semi- or fully automatic hydraulic linkage couplers and plain or cutaway discs. A semi-mounting kit, designed to reduce the load on the hydraulic linkage of lower horse-power tractors, was available for the HR42. An optional articulated top link arrangement for the HR44 made it possible to trail the discs when in work and to lift them on the hydraulic linkage for headland turns and transport.

An improved HR46A series of trailed A-type disc harrows for tractors up to 200 hp and with working widths of 10ft, 12ft 10in and 15ft 9in was launched in 1984. They had heavy box-section frames, hydraulically operated depth/transport wheels and the disc spindles ran on self-aligning ball bearings. The gang angle was adjusted with a turnbuckle and the wider models had hydraulically folded wings.

Steerage Hoes

The practice of horse-hoeing to clean the ground and stir the soil between the straight rows of growing crops had gained popularity through the eighteenth and nineteenth centuries once the use of the horse-drawn seed drill had been accepted as an improvement on broadcasting seed by hand. The hoes were usually made by a village blacksmith or agricultural engineer, and the wrought-iron horse hoes made by Ransomes in the late 1800s were little more than a horse-drawn scuffler with a singe front wheel and steering handles. Some horse hoes had small harrows following the hoe blades to stir the displaced weeds. The hoe blades could usually be set to work in rows between 14 and 24in apart.

The Howard Adjustahoe was added to Ransomes' implement range in 1932.

J. & F. Howard Ltd. were another company making horse hoes in the late 1800s. The Howard Adjustahoe steerage horse hoe was among the implements listed in their catalogue when Ransomes acquired the Bedford company in 1932. It was made in various sizes and, depending on model, could be set to hoe up to eleven rows of corn or three rows of roots at a time

Attachments for the C60 toolbar included a set of hoe blades, steering mechanism and a seat.

Attachments for the C60 toolbar designed for the E27N Fordson Major included ridging bodies and a steerage hoe.

The toolbar, with hoe blades spaced to suit row width and mounted on an E27N Fordson Major hydraulic linkage, was steered with a tiller handle by a second operator riding on a seat above an adjustable depth wheel. Hoe blades and a steering mechanism could also be used on the C66 toolbar made for Ford 8N and Ferguson TE 20 tractors. Similar steerage hoe attachments were also made for C72, C73 and C74 toolbars for the Ford Dexta and Major. The four-, five- or six-row TCR 1004 steerage hoe appeared in the late fifties. The independently mounted spring-loaded hoe units with individual adjustable depth wheels were spaced along a single square-section toolbar. Although Ransomes discontinued their steerage hoes the late sixties, the FR Cropguard Cleanrow chemical inter-row hoe (page 128) remained in production until the early seventies.

A steerage hoe was one of the attachments made for the FR C73 toolbar.

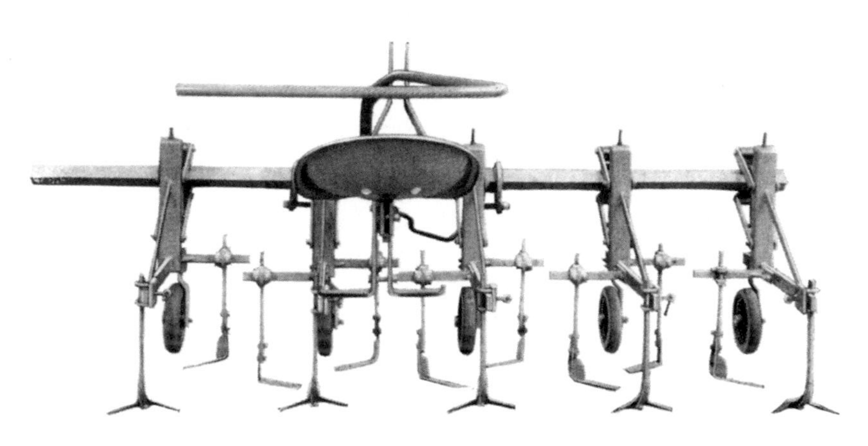

Sales literature explained that the TCR 1004 steerage hoe gave the operator a good view of the rows ahead. The slightest deviation could be followed with an accuracy that cut out the weeds without damaging the crop.

Mole Drainers and Subsoilers

By the mid-1800s farmers were already aware of the problems caused by soil compaction and understood how important it was to break up plough pans to improve drainage. Horses were the only source of power on most farms and Ransomes' catalogues in the 1850s illustrated various subsoiling attachments for horse ploughs and the RNS subsoiling plough. With a pair of Ransomes patent grubber tines bolted on to the plough beam behind the body, farmers were able to plough and subsoil the land at the same time. The catalogue explained that the grubber tines could be attached in a few minutes and removed just as quickly when the plough was required for ordinary work.

Another Ransomes design used a single subsoiler tine bolted to the beam behind the mouldboard. Three or four horses were needed to pull the plough when using this tine which, it was claimed, broke up the furrow bottom to a depth of 12in. A lever was provided to lift the tine out of work at the headland and a chain, linked to the front of the plough, was used to stop the tine swinging backwards out of work.

The RNS subsoiling plough was based on the Newcastle pattern plough. It had a narrow wrought-iron body designed to disturb subsoil up to a maximum depth of 18in. The handles were offset to the left allowing the ploughman to walk on unploughed land and, when conditions allowed, he could attach a set of steel knives to the plough beam to further pulverise the upper layer of soil.

Subsoiling tines were also used on some mounted and trailed ploughs. Two-furrow ploughs were normally used with the front body removed and a subsoiling tine bolted to the beam and held vertical by a heavy chain clamped to the plough frame. Subsoiling attachments were available for several ploughs over the years including the No.9 and No. 10 R.S.L.D., R.S.L.M., R.D.S.-T.C.P., Duotrac, Jumbotrac and the TS 55 mounted with a three-point linkage.

Tractor-drawn subsoilers and mole drainers were in production at the Orwell Works by the late 1920s.

The single-leg C8 self-lift model with a large-diameter disc coulter was used primarily as a mole drainer but could be converted to a subsoiler or heavy cultivator. The C8 had a maximum working depth of 20in and was said to be a suitable load for a light tractor. The heavier single-leg C1 subsoiler was similar to the C8 but required a medium-size crawler tractor. When it was set up for mole draining the C1 could also be used to lay underground cables.

Ransomes' No.7 tractor mole drainer, made in the 1920s and 1930s, was a different sort of design. Resembling a sledge, it had a disc coulter at the front and the mole was attached to an adjustable depth blade that could be reversed when worn. A handle at the rear controlled the implement when entering or leaving work and when the lever was pulled over sideways the mole automatically rose to the surface. The drainer was hitched to the tractor with a chain connected to a shackle at the front of the implement.

Provision was also made for the No.7 mole drainer to be pulled through the ground by a wire rope from the winch on a traction engine or farm tractor.

When a significant modification was made to an implement a new letter of the alphabet was added to model number. Thus by 1955 the C1 and the C8 subsoilers had graduated to the C1F and C8B. The C8C was discontinued in 1960 and the C1F subsoiler was dropped in 1965 but, as Ransomes built implements in large batches in those days, the last C1F was not sold until 1973. The C59, introduced in 1950 for high-powered crawler tractors, was listed as the C59B in 1955 when Ransomes added a trenching body, with adjustable mouldboards for making trenches 30 to 40in wide and up to 24in deep, to its optional equipment for subsoilers. Other accessories for the C1, C8 and C59 included a mole-draining attachment, large ridging bodies and heavy-duty cultivator tines.

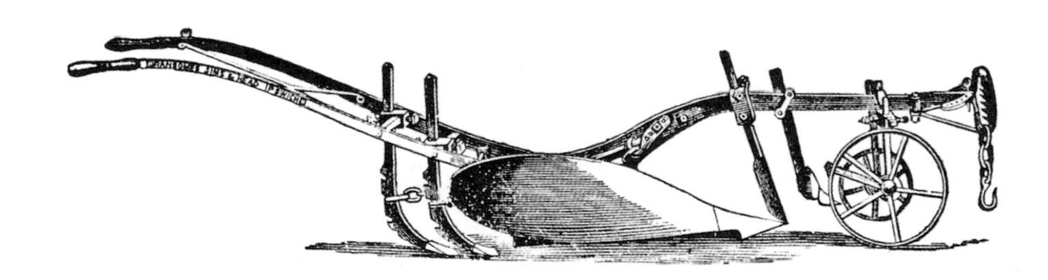

The patent grubbers on this Newcastle plough could be 'set just deep enough to break up a hard pan or pulverise the soil to a depth of 10in from the surface'.

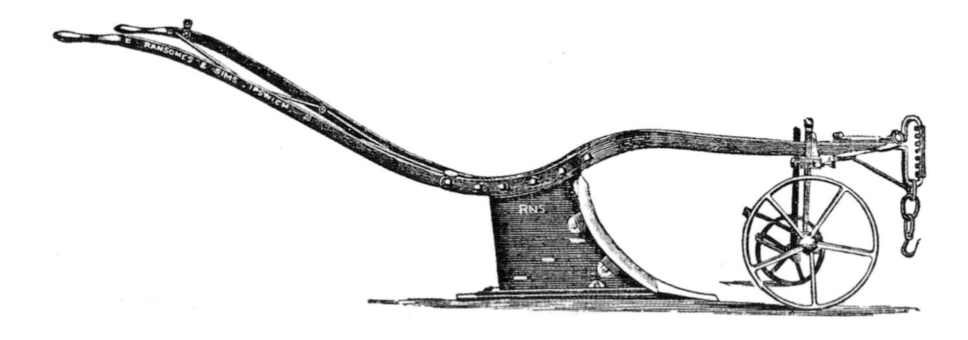

The RNS subsoil plough was used behind a mouldboard plough to break up the subsoil.

Ransomes' catalogue for 1932 explained that the arched beam on the No.7 tractor mole drainer made it ideal for working on rough grassland 'for the reason that if rubbish collects in front of the blade it can be removed without stopping work'.

The C1E subsoiler had a maximum working depth of 26in.

The C1 was discontinued in the late 1950s when Ransomes introduced the more efficient FR C75 mounted subsoiler with category II linkage for the Fordson Major and similar tractors. The C8 and C59 cultivators together with subsoiler, trencher and ridger conversion kits remained in production until the mid-1960s. The single-leg FR C85 subsoiler for the Ford 4000 and 5000 tractors replaced the C75 in 1966. It had a maximum working depth of 24 in, whereas the twin-leg FR C83 introduced in 1968, could only work to a depth of 15in. The C85, twin-leg C83 and the twin- or triple-leg C90 subsoiler version of the C90 heavy cultivator remained in production until the early 1980s.

The C94 mounted mole drainer appeared in 1980. Its leg and disc coulter were attached to a 12ft long twin-beam sledge, giving it a similar appearance to the No.7 drainer made in the 1930s. When the C94 was set into work the front end of its sledge was tilted downward by a patent headstock to give rapid penetration of the mole to its maximum working depth of 30in. The headstock was also designed to lift the front end of the sledge at the end of the run to take the mole out of work with minimum disturbance to the surface of the soil.

Launched in 1982, the C96 Subtiller was the first of a new generation of subsoilers with narrow legs and a wing share at each side of the point. There were two models: one had a 9ft 6in frame for five Subtiller legs or six heavy-duty cultivating tines while the 13ft 6in model was used with seven Subtiller legs or eight cultivating tines. Three Subtiller legs and a rear crumbler roll could also be bolted on to Ransomes' C83 and C90 toolbars.

The C1A subsoiler, also used as a five-tine cultivator or a mole drainer, broke up the subsoil to a depth of 22 inches and up to 20 inches deep when mole draining.

The front end of the beam was raised before the rear when the C94 mole drainer was lifted from work, and the front end was lowered first to draw the mole back into work at the start of the next run.

The C90 Subtiller with a maximum working depth of 16 inches was used to prepare land for direct drilling or minimal cultivation techniques.

The C90 heavy-duty toolbar was used for subsoiling, mole draining and heavy cultivating.

Rolls

Horse-drawn Cambridge, flat and Crosskill pattern rolls were made at the Orwell works in the 1850s. R. Hunt & Co. who were established at Earls Colne in Essex in 1824 manufactured a variety of farm implements at their Atlas Works including corn mills, horse works, thrashing machines, hay rakes and horse-drawn rolls, and by the 1920s they were making tractor-drawn rolls. Sir Reuben Hunt became chairman of Ransomes, Sims & Jefferies in 1951 and within a year or two tractor-drawn Cambridge and flat rolls were added to Ransomes' product range.

Ransomes' 1954 catalogue of farm implements included illustrations of Hunt-pattern single Cambridge rolls and steel-cylinder flat rolls from 7ft wide together with sets of three expanding or gang rolls. The catalogue explained that 'in the case of expanding rolls the two outer sections could be arranged in tandem for transport purposes.' The Cambridge-pattern expanding rolls could be supplied in 14, 15 and 16ft working widths and the gangs of light flat rolls were 13ft 6in and 17ft 4in wide.

Hunt-pattern heavy flat rolls made up with a series of 6in wide cast-iron sections were added 1956. The 8ft wide roll with 20in diameter cast rings tipped the scales at 15cwt and the 16ft roll with 24in diameter rings weighed over 2 tons. The maximum gang width

for the three types of roll was rationalised at 16ft in the early 1960s but within a couple of years gang width was increased to 24ft to match the ever increasing power of farm tractors. Ransomes stopped selling rolls in the early 1970s.

Ransomes introduced a range of furrow presses for the three- to six-furrow TSR 300 series reversible ploughs in 1985. Two rows of 26in diameter rings were carried on square axles mounted on heavy-duty sealed bearings. An optional weight tray was available for use on heavy land and a three-point linkage was also available for transport purposes.

Power Harrows

Responding to the popularity of power harrows in the early 1980s, Ransomes chose the 1985 Smithfield Show to launch a range of Italian-built rotary power harrows. There were three models with a two-speed gearbox and the 2.5, 3 and 3.5m harrows could be used on tractors with 540 or 1,000 rpm power shafts. Optional

The 1960s range of Ransomes Hunt Land rolls.

Furrow presses, originally used with horse-drawn and trailed tractor ploughs, were fashionable again in the early 1980s and Ransomes launched a new range of presses at the 1985 Smithfield Show.

equipment included an open-cage roller, packer roller and three-point linkage drill hitch.

New models of the Agrolux Ransomes power harrow with 2.5 to 4m working widths for 75 to 170hp tractors appeared in 1988. The narrower models could be used with a 540 or 1,000rpm power take-off but the 16 rotor, 4m harrow required a 1,000rpm power take-off shaft. There was a choice of a rear crumbler or packer roller, and optional accessories for the SKA and heavy-duty RKA harrows included mechanical or hydraulic three-point linkage drill hitches and a rear power take-off shaft.

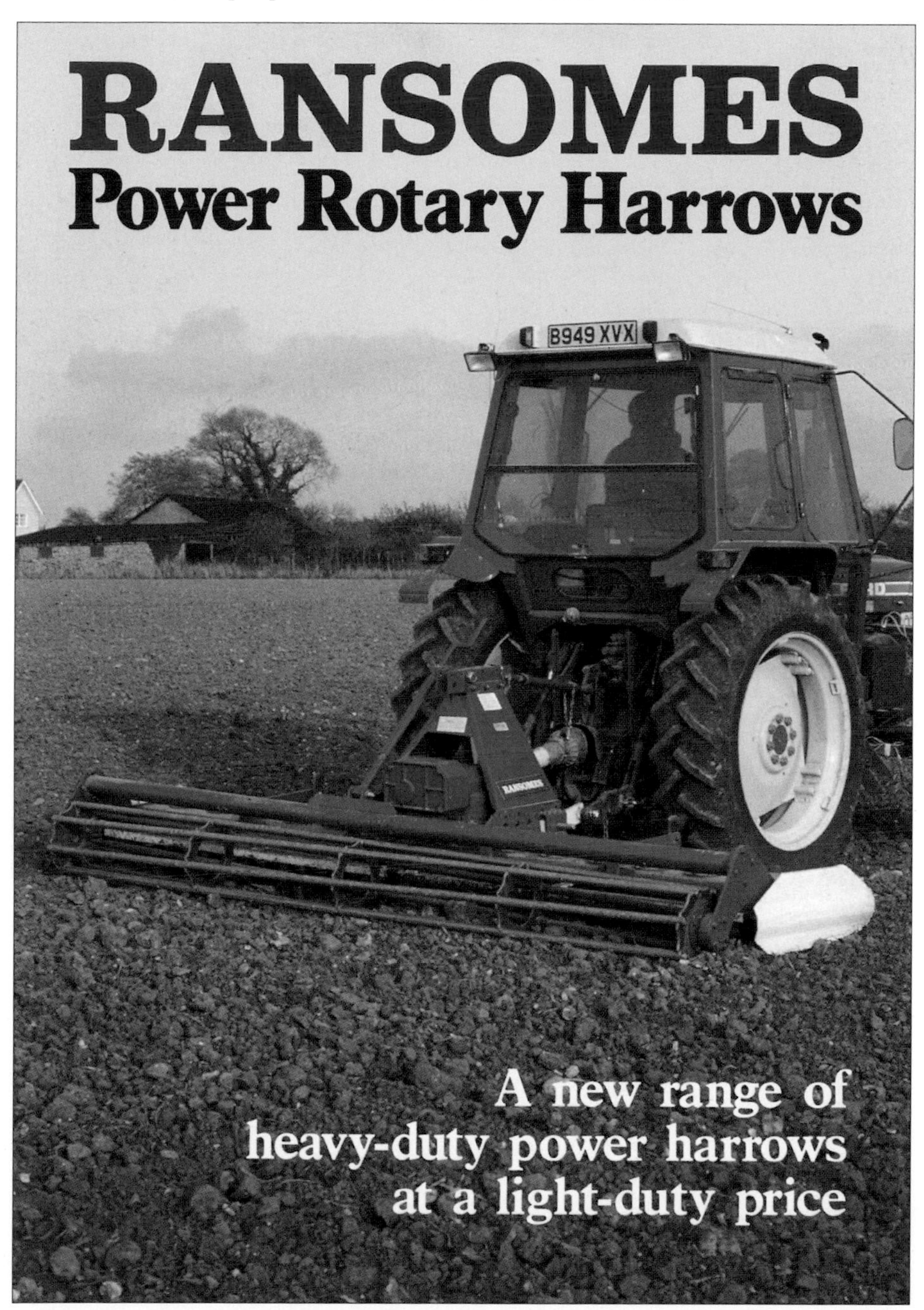

Ransomes' power harrows were made in Italy.

Chapter 4

Thrashers, Combines, Balers and Driers

Thrashing Machines

Thrashing may be spelt with an 'e' or an 'a' but it was invariably written as thrashing in Ransomes' sales and service literature.

Grain was thrashed with wooden flails until well into the 1800s when a crude form of hand-operated thrashing machine gradually came into use. James Allen Ransome exhibited a four-man hand-operated machine at the Royal Agricultural Society of England Show at Liverpool in 1841 and Ransomes used a self-moving portable steam engine to drive one of their thrashing machines at the following year's Royal Show at Bristol. The thrashing machine was mounted on a platform attached to the engine, the platform being lowered to the ground for use and then raised for transport.

In the 1800s only wealthy estate owners could afford powered thrashing sets, and the less affluent had to settle for a thrashing machine driven by a team of horses. Farmers with a small amount of grain to thrash could still buy a hand-operated machine in the early 1900s. A four-horse portable thrasher that weighed 39½cwt and cost £71 was listed in Ransomes & Sims' catalogue for 1862. The thrashing machine, horse gear and drive shafts were transported on a horse-drawn frame, and once the outfit was brought to the stack or barn it had to be unpacked and set out before thrashing could be started.

Traction engines were used to haul and drive thrashing machines by the late 1860s when the Royal Agricultural Society of England organised a series of thrashing trials at various locations in the country. Judges measured the power required by the machines and assessed the quality of the thrashed grain. The last great Thrashing Machine trials, held at Cardiff in

A hand-operated thrashing machine made by J. R. & A. Ransome in the 1840s.

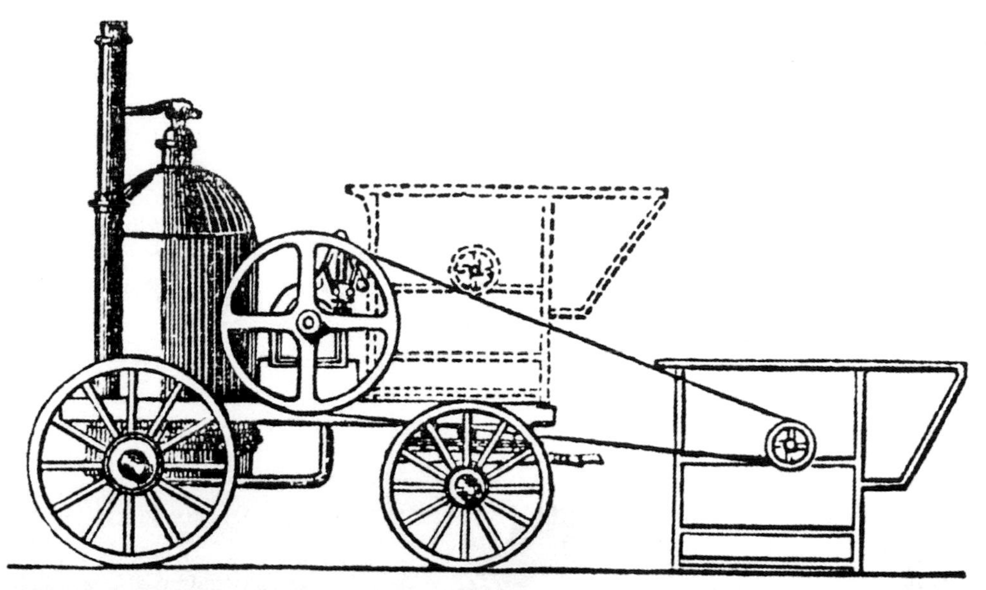

Ransomes & May portable steam engines were used in the mid-1840s to drive and transport a small mounted thrashing machine.

An early-1860s four-horse portable thrashing machine, packed for travelling (as above) and as used for work.

1872, attracted an entry of twenty-nine machines from thirteen manufacturers and of those Ransomes, Sims & Jefferies were the only one to be awarded two first prizes of £20 for their machines.

Seven horses were required to drive this thrasher which cost £100 on steel wheels. Wooden wheels added £8 to the price.

Single-blast, double-blast and double-blast finishing versions of Ransomes' large and lightweight thrashing machines were made at the Orwell works in the late 1800s. The 'blast' referred to the type of fan used to separate chaff and other waste matter from the grain. Class A, B and C versions of their large and lightweight machines were included in Ransomes' catalogue for 1886. The large machines had 48, 54 or 60in thrashing cylinder or drum widths; lightweight models, for small farms, were made with 36, 42, 48, 54 or 60in wide drums. Class A thrashers were double-blast finishing models with two dressing shoes and a patent adjustable screen 'to separate the grain into various qualities' and deliver 'a uniform sample of the best kind for the miller'.

Class B double-blast thrashers and Class C single-blast machines, mainly built for the export market, lacked this screen. If a clean sample was required it had to be dressed in the barn.

Catalogues recommended the use of double-blast machines in foreign countries where a large proportion of the grain crop was exported and it was suggested that the rather basic single-blast thrasher was the better choice in countries where 'skilled labour was scarce and repairs difficult to be effected'.

In 1885 Ransomes' cheapest model, a lightweight Class C thrasher with a 36in drum, was priced at £85, while their most expensive 60in Class A machine cost £150. Power requirement varied from 3nhp (nominal horse power, page 83), for the smallest model to 10nhp for the largest machine. Thrasher output depended on drum width, a typical 54in machine being able to thrash between 1,200 and 1,500 sheaves in an hour.

The Wilder patent self-feeder for steam thrashing was described in the 1886 Ransomes catalogue as a self-acting

Ransomes, Head & Jefferies were making double-blast finishing thrashing machines in the 1860s.

A page from the Ransomes & Sims Catalogue of Agricultural Implements and Machines for 1862.

safety-feeding apparatus. A new optional patent drum safety guard was available to 'make it almost impossible for anyone to fall into the drum of the machine'.

Ransomes exported thrashing machines to many different countries in the latter part of the nineteenth century. Most of them had a rasp-bar thrashing cylinder but machines with an American-type peg drum and concave were used for thrashing rice. Exports from the Orwell Works during this period included 143 thrashers and 137 Colonial traction engines to Argentina in 1892/93 and 338 thrashing sets to Russia in 1908. Another fifty traction engines and thrashers were sent to Turkey in 1914 but, because of the outbreak of World War I, payment was never received.

The Leviathan thrashing machine was introduced in 1892.

The AM medium thrashing machine had the capacity to thrash up to a thousand acres of cereal crops in a year. The optional cavings blower on this machine blew the cavings or chaff to a distance of up to sixty feet. The last Ransomes thrashing machines were made in the mid-1950s.

There were no major changes in thrashing machine design over the years. The introduction of optional pneumatic-tyred wheels was an advantage to contractors who were using farm tractors to haul and drive their thrashing sets. Although combine harvesters were already replacing the thrashing machine on some large farms in the mid-1930s, small-scale production of thrashing machines continued at Ipswich until the mid-1950s.

Various modifications were made to the Class A heavy-duty thrasher during its sixty-year production run. Designed for use with steam engines, and later with high-powered farm tractors, the Class A thrashing machine with its 54in drum was recommended for use by contractors and large farming estates. The AM medium-class thrasher with a 48 or

The Invincible clover huller thrashed up to two tons of clover seed in a day.

54in wide drum and suitable for 20 to 25hp tractors was aimed at farmers doing some contract work to increase their income. The Ransomes Tractor Thrasher with a 42 or 54in drum was a full-sized lightweight machine for private use and recommended for farms where weight considerations were a deciding factor or where the terrain was ' heavy-natured country or downland.'

Clover was said to be difficult to thrash without a very severe rubbing and a Ransomes clover huller provided an 'efficient and economical method of dealing with moderate quantities of clover'. A clover huller consisted of a cylinder rotating in a wire cage placed on top of a thrashing machine. An aspirator fan lifted the clover heads or knobs, already thrashed by the main drum, from the lower sieve up to the huller. Any seed left in the heads was removed by the rotating cylinder and conveyed to the dresser while the empty seed heads and chaff returned to the main sieves where they were blown out of the machine by the fan.

A sheaf elevator, which could be used on either side of the drum, eliminated the need to pitch the sheaves from a wagon on to the thrashing deck when the crop was thrashed at harvest time.

Farmers and contractors handling large quantities of clover, lucerne, vetches, etc., were advised to buy a Ransomes Invincible clover huller. It could thrash up to 40cwt of clover seed in a day and was similar to a thrashing machine with the huller unit housed within the body of the machine. Potential customers were advised that the Invincible would produce a perfect sample of seed with 'the clover being treated in a scientific manner from the moment it enters the machine until the delivery of seed ready for the market.'

Ransomes' straw elevators 'carried thrashed straw to a height of 20ft and saved the labour of three men'. This model on iron wheels cost £47 in the late 1880s.

A steam engine or horse gear was used to drive Ransomes' stacking machines.

A straw elevator or pitcher, and on some occasions a chaff cutter, were driven by belt from the thrashing drum. The elevator was towed from farm to farm behind the drum and when the chaff cutter was required it was usually hitched up to the elevator. Ransomes, Head & Jefferies were making straw elevators and stacking machines at the Orwell Works in the late 1800s. The straw elevator could be set to deliver straw from the straw walkers in any direction from the thrashing machine. Ransomes built two sizes of elevator, 20 or 22ft in length, both with a 4ft wide trough. They were said to save the labour of three men.

Stacking machines could either be used with a thrashing drum, or for elevating sheaves or loose hay when building a stack in the field or stack yard. They could be belt-driven from the thrasher or from a portable steam engine, or else they could be more cheaply powered by horse or pony gear at hay time and harvest. The horse gear consisted of a long pole attached to a large-diameter gear wheel, which transmitted power through a system of shafts to the elevator or other machine. One or two horses were harnessed to the pole and when they walked round at normal speed the output shaft turned at approximately 400 rpm.

The Ransomes & Sims catalogue for 1862 explained that a lad could produce six bushels of chaff in an hour with the No.14 hand-powered chaff cutter.

Chaff Cutters

Ransomes made various types of hand-operated and power-driven chaff cutters in the 1830s *The Journal of the English Agricultural Society* praised the No.12 chaff cutter on Ransomes' stand at the 1839 Royal Show at Oxford. It could be driven by horse, water or steam power and was described as the 'largest and most powerful yet seen and remarkable for its equable slicing cut made by the two knives, each 3ft long, on a flywheel'. Straw was moved towards the knives by a crank-operated press board and when it was driven by two horses the No.12 cut chaff between $\frac{3}{8}$ and $1\frac{1}{2}$in long at a rate of 10cwt an hour. At Oxford Ransomes also displayed a hand machine with a screw mechanism to move the straw to a single knife that cut chaff at a fixed length.

Ransomes & Sims' 1862 catalogue of agricultural implements and machines

Ransomes recommended the purchase of a spare knife wheel when buying their patent portable steam-powered chaff engine so that 'one set of knives could be sharpened without removing them from the wheel while the other is at work.'

illustrated the improved hand-powered No.14 chaff cutter. It had two knives, a capacity of up to 6 bushels of chaff an hour and cost £3 15s. The heavier No.16 iron chaff cutter for the export market was £4 4s. The same catalogue included the No.5 steam-power chaff cutter with rise and fall feed rollers and an hourly output of 30 to 35cwt of ½in long chaff.

An improved chaff cutter or chaff engine with a patent safety feeder to receive straw direct from a thrashing machine was made at Ipswich in the late 1880s. It had a chaff sifter to remove the dust and an attachment to bag the chaff off for animal feed. When the chaff cutter was used during thrashing it was driven by belt from the drum; alternatively, when it was used to cut chaff from previously thrashed straw, the belt was driven from a portable steam engine flywheel. Chaff could be cut in lengths of 3/16, ⅜ or 1in; the dust was sifted out and the chaff was delivered into bags, ' in a fit state for storage with great rapidity and at small cost '.

The catalogue explained that 'the machine was furnished with a patent lever for instantly stopping the roller in case the hand of the feeder should be drawn in and the few cog wheels on the machine are cased over so that no danger can arise from them'.

Combine Harvesters

Ransomes were winning medals for their thrashing machines at agricultural shows in the 1840s but more than a hundred years were to pass before they produced their first combine harvester. A false start was made in 1946 when, in an attempt to meet a request from a Yorkshire farmer, Ransomes' engineers converted a standard thrashing machine into a trailed combine harvester. A pick-up cylinder and an elevator were attached to the front of the thrasher to collect and convey a previously cut swath to the self-feeder at the top of the machine. An engine was mounted on top of the drum, a platform was provided for bagging-off the grain and the chaff was returned to the ground. Although the machine worked well enough it was rather heavy and required that the crop should be cut and swathed before it could be used. Ransomes decided to continue making thrashing machines and kept out of the rapidly growing combine market until 1953 when, realising that thrashing machines had had their day, they introduced the Bolinder Munktell MST 42 to British farmers.

Ransomes' engineers experimented in the mid-1940s with a trailed combine harvester. It was built round a thrashing machine with the pick-up borrowed from a straw trusser, and power came from an engine mounted on top of the drum.

Ransomes tried their hands at most things including sail reapers. This 1860 model has a heavy cast-iron driving wheel with cast teeth. The man is probably waiting for a fresh team of horses.

The company announced that supplies of the Swedish-built 4ft cut power-driven trailed combine would be imported in order to meet the initial needs of their customers and that it would be made under licence at Ipswich for the 1954 harvest. A 25 to 28hp tractor was recommended for the MST 42 combine that had a bagging-off platform 'of spacious dimensions', an optional pick-up attachment for harvesting previously windrowed crops and a 42in wide thrashing drum. Ransomes preferred the term 'drum' to 'thrashing cylinder' for their thrashers, and applied the same usage to most of their combine harvesters.

Bolinder Munktell of Sweden made the MST 42 combines sold by Ransomes for the 1953 harvest.

The 1956 Agricultural Machinery Census recorded about 40,000 trailed and self-propelled combine harvesters and 12,300 thrashing machines in England and Wales. By 1959 these figures had changed to 48,000 and 10,500 respectively, though it is doubtful that many of the thrashers were still in use.

The Ransomes MST 56 power take-off driven combine cost £685 when it was introduced in 1956.

The 5ft cut MST 56, with thrashing and separating mechanisms similar to those of the MST 42 appeared 1959. The new model could be ordered with a bagging-off platform or a grain tank and optional equipment included a Reynolds or a Wicksteed pick-up reel and a rotary screen on the bagger model 'to eliminate weed seeds'. An MST combine with a Ford Zephyr engine was built at Ipswich in an attempt to compete with the engine-driven International B 64 but the project was dropped at the prototype stage.

The price of the 10ft cut 902 combine harvester started at £2,370 when it was launched in 1958.

Although Ransomes took a close look at Bolinder Munktell's self-propelled combine, they decided to

The 801 combine harvester had an output of about three tons an hour.

design their own harvester and the pilot model of the first Ipswich-built self-propelled 902 combine harvester appeared at the 1956 Smithfield Show. Selected farmers tested twenty pre-production machines during the 1957 harvest and a production model was exhibited at the 1958 Royal Show. The 902 was a straight-through 6 tons per hour combine with a 10 or 12ft cutter bar, 24in diameter and 39in wide drum, four straw walkers and rotary screen on the bagger model. The drum was chain driven and different sized sprockets were used to alter its speed for different

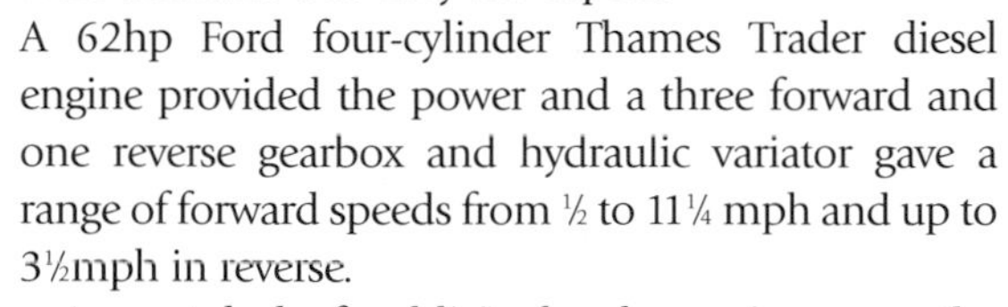

crops. Wider 14 and 16ft tables were available but only for export. A 62hp Ford four-cylinder Thames Trader diesel engine provided the power and a three forward and one reverse gearbox and hydraulic variator gave a range of forward speeds from ½ to 11¼ mph and up to 3½mph in reverse.

A great deal of publicity has been given over the years to the teams of Massey-Harris and John Deere combine harvesters which combine their way across America. A fleet of Ransomes' 902 combines achieved a similar feat by working their way across Syria in 1962. The harvest was started in May and took seventy-eight days to complete in late July. During that time Ransomes' service engineer Doug Jaye drove his 902 a distance of approximately 3,375 miles and combined nearly 5,800 acres.

Hydraulic table lift and disc brakes were features of the 1001 and 801 combine harvesters.

The lack of demand for trailed combines led to the demise of the MST 56 in 1961. Although the 902 was still in production, Ransomes decided that it was time to introduce a new self-propelled machine. In those days it was usual to build a prototype in the Experimental Department (page 38), so Ransomes' engineers were presented with an engine and various parts from the 902 and told to get on with the job. Once built, the prototype was tested in the field and the necessary modifications were made before it was passed to the drawing office. They in turn prepared the working drawings needed for the foundry, machine shops, welding bays and production line.

The result of their combined endeavours was the bagger and tanker versions of the self-propelled Ransomes 801 launched in time for the 1963 harvest. Sales literature for the 8ft cut combine explained that with its 8ft 6in transport width the 801could be driven 'along narrow lanes and tracks less accessible to other combines in the same class.' Specifications

The Gentleman Apprentice

There were two types of apprenticeship at Ransomes: craft apprentices learned the skills of a particular department and gentlemen or special apprentices were given a broad training to prepare them for management within the company. Following a two-year spell on his father's farm at Braintree in Essex, Fred Dyer was taken on as a special apprentice in September 1933.

It was a complete change of life. The works hooter, or 'Bull' as it was generally known, sounded at 7.30 am when the gates were shut and the workers had to be at their benches. The gates were opened again at 7.35 am and latecomers lost fifteen minutes pay for that day. Like all new special apprentices, Fred was given one of the white smocks that led to the nickname of 'ice-cream men' used by workers on the shop floor. Fred was given a 'home' to keep the smock in the foreman's office, a room that had its own washing facilities and was to be Fred's base for the next three years until he was moved on to the Plough Works.

Taught by highly skilled men in the engine fitting shop, Fred worked on the last steam engines made at the Orwell Works. While in this department for nine months, he polished steam-engine connecting rods, hand-chipped cylinder blocks and from time to time worked on trolleybuses, corn mills and electric trucks.

Ransomes' apprentices were expected to endure various 'leg-pulls' such as being sent to the stores for a 'long weight' only to find half an hour later that it was spelt 'wait'. There were times when snow fell through the fitting shop's leaky roof on to the benches. Heating was poor with only a few coke fires and steam radiators. In summer a sudden thunderstorm would flood the shop so that the wood block floor swelled up. Apprentices sometimes made things for their own use such as tee squares, drawing boards and toolboxes. Known as 'Government Work', some items were used for work purposes while others were smuggled through the works gate.

The foundry and the pattern shop were next on Fred's list. The foundry was the dirtiest place imaginable with an atmosphere of dust, smoke and smell. Women staffed the core shop and apprentices rarely went into this shop unescorted! The pattern shop was built above the railway on Ransomes' quay where pig iron ingots were unloaded for the foundry. Mahogany and pine were used by the pattern makers who measured their work with a contraction rule to suit the type of metal used to make the casting.

The Forge Shop was the next port of call and here gentleman apprentices learned how to bend hot steel bars. Other skills gained included using steam stamps to make steel plough shares, bulldozers to bend plough beams and hydraulic presses to form thrasher shaker cranks.

Helping to build a variety of thrashers, twin-drum clover hullers and a few balers for both home and export markets was an interesting experience during the three months Fred spent in the Thrasher Works. It had its own steam-driven band saw, partly fuelled with sawdust and wood shavings from wood-machining tools. The saw was used to convert ash and oak tree trunks into suitable sized pieces for thrasher parts, potato-digger poles, beams for wooden ploughs, lawn mower rollers and many other items.

Most of the final training period was spent in the Plough Works with its own machine shop, progress department and design office where Fred learned how much time and effort was put into the design and testing of new implements. The last few months of apprenticeship were spent as assistant to the Works Manager, who sent him on visits to Ransomes dealers. Other tasks included organising stores department staff sorting a five-ton heap of new bolts and dealing with a huge stock of obsolete horse-plough frames which it was agreed would be sold for scrap.

On completion of his gentleman's apprenticeship Fred Dyer, as Assistant Works Manager, went on various overseas tours to South Africa and other countries before being appointed Plough Works Manager in 1943, and within twenty years he was Manager of the Farm Machinery Works at Nacton. On his retirement in 1979, Ransomes' house magazine commented that Fred Dyer must have been one of very few engineering works managers to have regularly worn breeches and a white smock.

Fred Dyer with his 1869 Ransomes, Sims & Head mill.

The 12ft cut Crusader (left) had an output of up to 7½ tons an hour and the 14ft cut Cavalier harvested up to 9 tons an hour.

included a 42hp Perkins engine, 36in drum and four straw walkers. The Handi-matic control system on the driving platform provided instant adjustment of the variable speed vee-belt drives to transmission, reel and thrashing drum.

The 10 and 12ft cut Ransomes 1001 tanker combine harvester cost from £2,695 when it superseded the 902 in 1964. In common with earlier self-propelled Ransomes combines the engine was mounted on the chassis for added stability when working on hillsides. The 6 tons an hour 1001 was a scaled up version of the 801. It had a 62½hp four-cylinder or optional 90hp six-cylinder Ford diesel engine and a top speed of 11 mph provided by a three-forward and reverse gearbox with a variator drive unit.

Ransomes' Research and Development Unit spent five years designing and building a new combine harvester and a prototype Cavalier was shown to farmers at the 1965 Smithfield Show. Twenty-five pre-production models were built in 1966: twenty-four were field tested by farmers in various parts of the country and the twenty-fifth took part in the Ransomes Cavalier roadshow. The plan was for Bruce Dawson, with John Horner and Neville Walsingham to demonstrate the combine on a field within sight of the English Channel and then drive it by road to finish up on a field overlooking the Forth Bridge. To aid farmers willing to host demonstrations en route, the Ford demonstration team at Basildon loaned a Thames Trader lorry to transport a Ford 5000 Selecto-speed tractor and trailer to be used to shift the grain from this new high-capacity combine. However, the demonstration team encountered bad weather and difficulties of a 'technical nature' when demonstrating the machine so they concentrated on solving the technical problems and got no nearer to the Forth Bridge than a farm in the Midlands.

Twin drums were a revolutionary feature of the Cavalier and Crusader combine harvesters. Hydrostatic steering and an automatic floating cutter bar were standard, and a cutting height indicator was included in the £3,900 price tag on the Cavalier.

One final batch of 1001 combines came off the production line late in 1966 followed by a group of Cavaliers built in time for the 1967 harvest. Advertised by Ransomes as a 'new concept in combine design', the 9 tons an hour Cavalier for the larger acreage had a unique twin-drum and concave thrashing system. Ransomes' engineers stepped up the thrashing capacity by installing a 45in wide and 8¾in diameter primary drum in front of the main drum. This new 'Early Action' feature separated out up to 25 per cent of the grain and provided an even flow of the crop to the 24in diameter main drum and concave to improve the final thrashing and separation of grain from the straw. A six-cylinder 90hp Ford diesel engine, mounted low down on the chassis, provided the power for the 12 and 14ft cut Cavalier. It had four straw walkers, hydrostatic steering, a top speed of 12mph, weighed 5 tons and it only took 70 seconds to empty the 70 bushel grain tank. The Cavalier was equipped with lights to indicate cutting height and an amber flashing beacon to alert a distant tractor driver when the grain tank was almost full.

The 7½ ton an hour Crusader, designed to replace the 801, made its debut at the 1967 Smithfield Show. The 10 or 12ft cut Crusader with a 74hp Perkins engine was a smaller version of the Cavalier with the same primary drum but with an 18in diameter main drum. The 10ft cut Crusader cost from £2,995 when it was launched but within five years the price had almost doubled to £5,390. Approved accessories for the Cavalier and Crusader included a Reynolds draper pick-up attachment, an all-weather cab made by Tractorvision at Wolverhampton and a trailer for the detachable 14ft cutting platform made by John Baker & Sons at Sheffield. In response to customer demand Ransomes added a 10ft table for the Cavalier and an 8ft 9in table for the Crusader to the list of optional equipment.

The Super Cavalier with a choice of 10, 12 and 14ft wide quick-release cutter bars cost from £6,780 in 1974 when it replaced the Cavalier. The concept of twin-drum thrashing disappeared from the British market when the last Super Cavalier was built at Ipswich in 1976 but within a decade other manufacturers had 're-invented' the principle of early-action thrashing with twin drums originally developed by Ransomes in the early sixties.

One hundred and thirty years of unbroken manufacture of thrashing machines and combine harvesters at Ipswich came to an end when the last Super Cavalier was driven off the Nacton Road production line in 1976.

Balers

In the 1880s Ransomes were making hand-operated straw trussers and 'straw trussing apparatus' for use with steam thrashing machines. The hand trusser consisted of a heavy length of timber with four vertical U-shaped yokes. Straw was placed between the arms of the yokes and compressed into a dense bundle by pulling the tops of the yoke arms together with cord and securing them while the truss was bound with twine.

The steam-driven trusser could be mounted on a pair of transport wheels and hitched to the thrasher or permanently suspended from the thrashing machine. It was chain-driven from the straw walker shaft and made trusses weighing up to 36lb. A catalogue at the time

Ruston & Hornsby were making suspended straw trussers for use with thrashing machines in the late 1880s, and Ransomes were still making the Hornsby No.4 trusser in the early 1950s.

The shafts on the wheeled version of the No.4 trusser were also used to guide the trusses on to the ground.

explained that the twine needed to tie up a ton of straw cost a shilling. This was no more than 'the cost of labour for making straw bands to bind the same weight of straw whilst the whole of the labour for trussing is saved'. Prices in 1886 were £47 10s for a double band trusser on two wheels complete with a pair of horse shafts or £37 10s when suspended on the thrashing machine. An alternative suspended model with a single knotter cost £27 10s.

Ruston & Hornsby had been making steam engines, thrashing machines and straw trussers at Lincoln since the 1850s and, following Ransomes' association with the Lincoln company in 1919, the Hornsby No.4 straw trusser appeared in Ransomes' thrashing machine catalogues. The No.4 trusser could be suspended on the thrashing machine or mounted on two wheels. The wheeled model was either towed from farm to farm behind a thrashing machine or fitted with shafts and moved around by a horse. The trailed model was suitable for use with any leading make of thrashing machine and made ready for work in a matter of minutes. There was no need to remove the horse shafts as they provided 'an easy means of descent' when loading the trusses on a farm wagon.

The suspended trusser could be set so that straw was trussed as it left the thrashing machine or lowered on its support chains, allowing the straw to bypass the trusser when it was being stacked loose. The No.4 was chain-driven from the straw walker shaft on the thrashing machine and could be supplied with one or two binder knotters With suitable straw it could tie bundles weighing 5 to 12lb, 12 to 25lb or 26 to 35lb but unless otherwise ordered the No.4 was sent out set to tie trusses weighing between 12 and 25lb.

Up to twenty tons of high-density wire-tied bales of hay or straw could be made in a day with the Ransomes stationary baler.

Wheeled and suspended No.4 trussers were still being made in the late 1940s and early 1950s when Ransomes were also manufacturing high-density hay and straw balers with some Ruston & Hornsby engineering in its pedigree. Straw was normally baled directly from the thrashing machine but when hay was being baled the crop was carried to the baler with a hay sweep or wagon and forked

Thrashing and baling in 1947 with a Ransomes AM thrasher and high-density baler.

by hand or mechanically elevated to a patent automatic feed conveyor. This fed the crop into the bale chamber by means of a series of reciprocating tines projecting through the floor of the feed hopper. The horse's-head-action feeder arm pushed the crop into the bale chamber where the heavy cast-iron ram, running on renewable tracks, formed the bale. Large needles were passed between each bale by hand, wire was threaded through the needles and the ends were twisted together. Wire-tied bales were never popular with livestock farmers and for this reason an optional twine-tying attachment was introduced in the late 1940s.

Ransomes acquired D. Lorant Ltd. at Hartspring Lane in Watford in 1951. Lorants had been importing Claas low-density stationary and pick-up balers since 1947 and selling them as Lorant balers. Ransomes, Sims & Jefferies [Watford] Ltd continued the baler business at Hartspring Lane and three pick-up and one stationary low-density balers were added to the Ransomes product range. The Model M driven by an air-cooled 8hp Armstrong Siddeley diesel engine made 39 x 18 x 24in bales and had a capacity of 3½ to 5 tons per hour depending on the type of material.

LORANTS

D. Lorant Ltd was formed in Hertfordshire in the mid-1940s when Mr Lorant started to import low-density balers from Claas in Germany. The balers were complete, apart from their wheels, which Lorants fitted in their small factory at Radlett.

An agreement between Claas and Lorants in 1948 led to the manufacture of the baler in the U.K. Machining and knotter assembly were carried out at a small factory at Letchmore Heath while all other assembly was done at Radlett. When Ransomes acquired the Lorant business in 1951, baler production had been moved to Hartspring Lane, Watford, the site of a larger factory that had been used as a test bed for Napier aeroplane engines during World War II. Ransomes sent Mr J. Cannings and Mr E Garrard to Watford as managing director and works manager respectively and continued to manufacture low-density balers there until production was transferred to Ipswich in 1956. The Lorant name was retained for a couple of years and then the balers were marketed under Ransomes' name.

The self-propelled Vibro Hoe (page 95) used for inter-row hoeing in horticultural crops was also made at Watford from 1954 until the factory closed in 1956, when production was transferred to Ipswich.

It appears that farmers in the 1950s had time to write testimonial letters to manufacturers. One such letter, published in Ransomes' sales literature, read, 'As you know, I am not easily satisfied with agricultural machines generally, and am apt to be critical both of design and workmanship. I am therefore very pleased to tell you that I have no fault to find with the model M baler supplied to me and have had absolutely no trouble or difficulty whatever during the baling of several acres of hay and straw.'

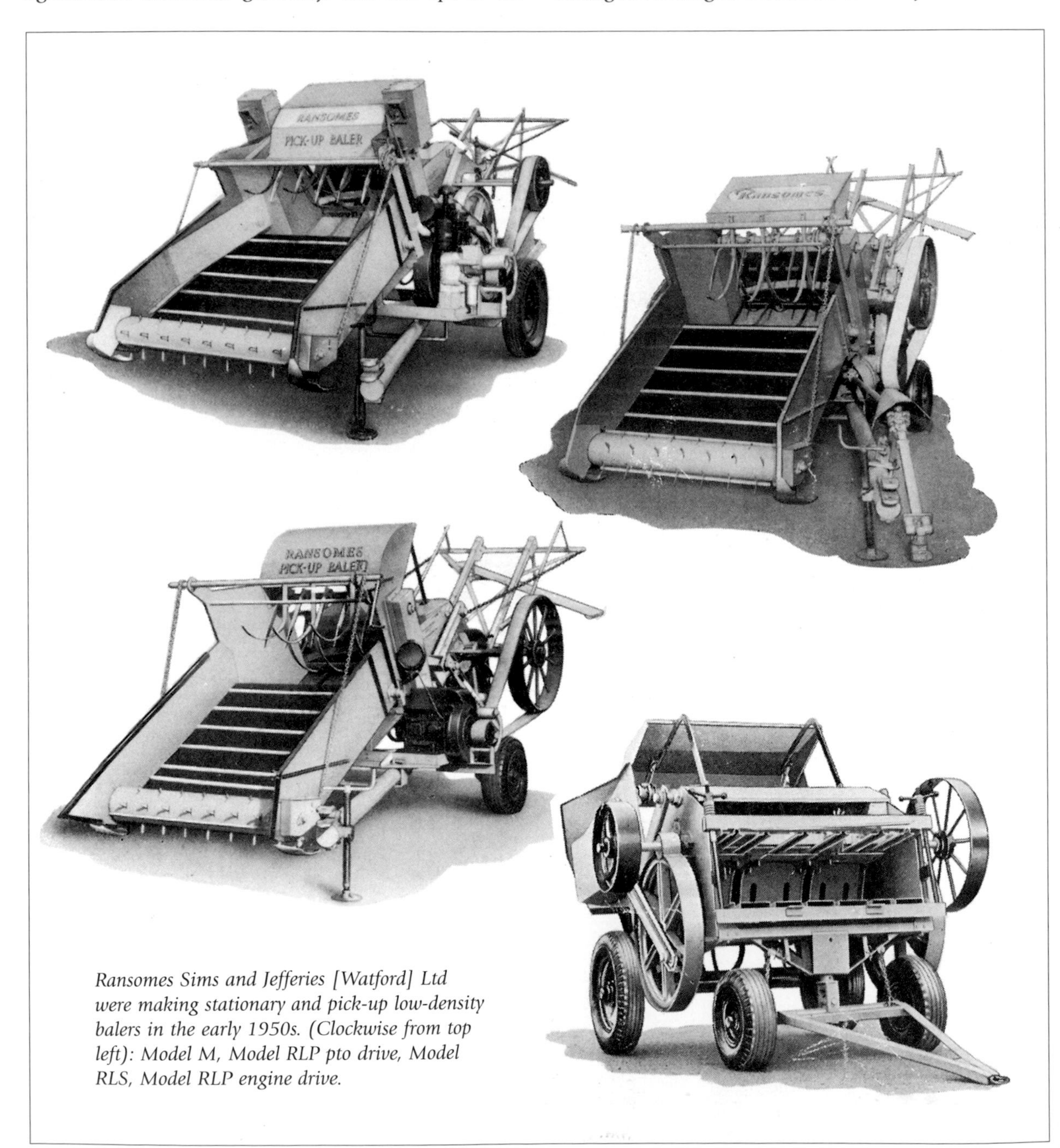

Ransomes Sims and Jefferies [Watford] Ltd were making stationary and pick-up low-density balers in the early 1950s. (Clockwise from top left): Model M, Model RLP pto drive, Model RLS, Model RLP engine drive.

Bales could be loaded directly on to a trailer towed behind a Ransomes pick-up baler.

The smaller 2 to 4 tons per hour Model RLP baler could be driven from the power take-off or by a 10hp twin-cylinder air-cooled Petter engine. Ransomes recommended an 18 to 20hp tractor for the power-driven baler. The belt-driven RLS stationary baler, mainly used by thrashing contractors, had a wider bale chamber with an output of 2 to 5 tons of 51 x 24 x 20in bales in an hour.

The RLS and engine driven RLP were discontinued in 1955 but Ransomes were still selling low-density balers in 1958 when the Model M cost £700 and the power-driven RLP was £500.

Driers

Ransomes introduced a general-purpose, horizontal crop drier in 1936. Made under licence from British Crop Driers Ltd, the two- and three-stage oil-fired

About 4 cwt of grass or 20 to 25 cwt of grain could be dried in an hour with the Ransomes BCD 9/4 oil-fired continuous drier.

continuous driers were suitable for grass and grain. As an alternative for grass only, solid fuel could be burned with the aid of an automatic stoker.

The oil-fired 27cwt/hr Model 187 gravity-flow drier was introduced in 1955 to meet an increasing demand from farmers changing from traditional harvesting methods to the combine harvester. The 187 used the rather high maximum drying temperature of 180° F and the warmed air was blown through the grain as it flowed down through the top section of the drier. Cold air was blown through the lower part of the drier to cool the grain for storage.

Within a few years the capacity of the Model 187 drier had fallen behind the output of the latest combine harvesters. Ransomes overcame the problem with the introduction of the 65 cwt/hr GF 195 and GF 196 oil-fired gravity-flow driers in 1960, and the following year the smaller 34cwt/hr GF 193 and GF 194 superseded the Model 187. Ransomes' grain drier output was calculated with the air heated to a temperature of 160° F to reduce grain moisture content from 21 per cent to 16 per cent. The GF 193 and GF 195 were for right-handed operation while the GF 194 and GF 196 were suitable for

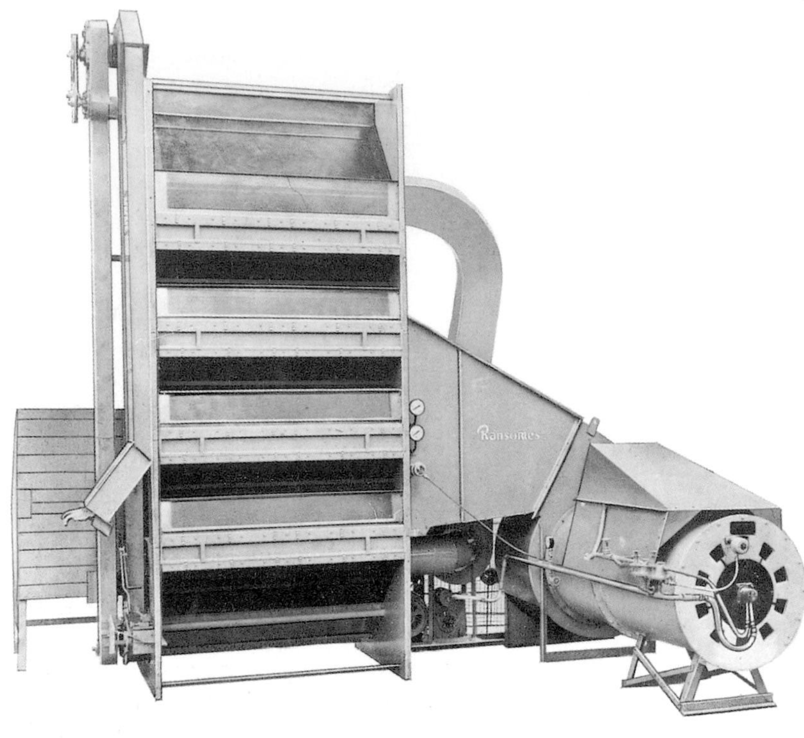

The Model 187 vertical gravity-flow grain drier with an output of 20 to 25 tons in a day could be installed in an existing building without providing any additional foundations.

There were no mechanical conveyors or other moving parts in the drying and cooling sections of the five and ten ton per hour Multivac driers.

installations where left-handed operation was required. Prices in 1962 started at £1,037 for the smaller model and from £1,229 for the GF 195/196. Special foundations were not required for these driers which had fully automatic controls, a flame failure device and a 10hp motor to drive the main fan.

The Multivac 5 and 10 tons/hr oil-fired continuous driers were launched in time for the 1966 harvest. Warm air followed by cold air was drawn through the grain by a suction air-flow system which dried and cooled the grain as it passed down through a honeycomb system of ducts to the discharge hopper. The 5 tons/hr drier with electric motor power totalling 25hp cost £2,665 in 1972 and the 10 tons/hr model with a total of 48½hp of electric motors was £4,165.

Although it was not one of their major products, Ransomes introduced a range of grain-handling equipment at the 1968 Royal Highland Show. Made for four or five years, the range included various sizes and lengths of grain elevator with capacities of 10, 15 and 20 tons/hr as well as single and double elevators in any length with outputs of up to 20 tons/hr. A variety of fittings and attachments was available, including a cut-off outlet for high-level conveyors which could be operated at ground level so that a catwalk was not needed.

1st JANUARY, 1951. ALL PREVIOUS LISTS CANCELLED. No. 15318H.

Prices of

Low res - needs scanning.

Ransomes

THRASHING MACHINES

Mounted on Iron Wheels and Pneumatic Tyred Wheels

Thrashers

Type	Price on Iron Wheels £ s. d.	Price on Pneumatics £ s. d.	Sizes of Pneumatic Tyres
A 54"	920 0 0	1,025 0 0	Front: 9.00 × 16 Rear: 9.00 × 16 H.S.
AM 54"	870 0 0	965 0 0	Front: 9.00 × 16 Rear: 9.00 × 16
AM 48"	840 0 0	935 0 0	Front: 9.00 × 16 Rear: 9.00 × 16
T 54"	820 0 0	885 0 0	Front: 8.00 × 19 Rear: 8.00 × 19 H.S.
T 42"	750 0 0	815 0 0	Front: 8.00 × 19 Rear: 8.00 × 19 H.S.
LAL 36"	635 0 0	700 0 0	Front: 7.00 × 19 Rear: 7.00 × 19
SA 27"	430 0 0	495 0 0	Front: 7.00 × 19 Rear: 7.00 × 19

Clover Huller

	£ s. d.	£ s. d.	
48"	743 0 0 (Chaff Bagger £30 0 0 extra)	808 0 0	Front: 8.00 × 19 Rear: 8.00 × 19 H.S.

Hay and Straw Baler

	£ s. d.	£ s. d.	
	562 0 0	622 0 0	Front: 7.50 × 10 Rear: 9.00 × 16

All prices are subject to alteration without notice and orders are accepted only at prices and terms ruling at date of despatch.

RANSOMES SIMS & JEFFERIES, LTD., ORWELL WORKS, IPSWICH

E.A. D. 511.

Chapter 5

Steam Engines and Tractors

The first Ransomes steam engine was exhibited at the Royal Agricultural Show held at Liverpool in 1841. It was a portable engine, essentially a stationary engine and vertical boiler mounted on a wheeled carriage, with shafts at the front so that it could be hauled from place to place by animal draught. The complete unit weighed 35cwt and was pulled by two horses, or more if the ground was soft.

Stationary engines had already been used to drive fixed thrashing machines, and Ransomes went on to link thrashing more closely with portable engines by mounting a small thrashing machine on a platform at the back. Originally operated by four men, the thrasher was driven by belt from the engine flywheel. The platform was lowered to ground level for use and then lifted again when the stack had been thrashed.

Ransomes made history when they exhibited a self-propelled version of the engine at the Royal Show held at Bristol in 1842, a machine that was probably the world's first self-propelled machine for agriculture. The front wheels were driven by chain from the crankshaft and it had a top speed of 6mph.

John Head, a leading figure in the development of steam power at Ransomes, developed the world's first practical self-moving road-going steam engine in 1849. Made at the Leeds railway foundry, the Farmer's Engine was designed for belt and haulage work. When used for belt work, the rear driving wheels were jacked off the ground and the chaff cutter or other machine was driven by belt from one of the wheels. Little more was heard of the Farmer's Engine but within a couple of years Ransomes were building steam engines at Ipswich with many of the features used on earlier models.

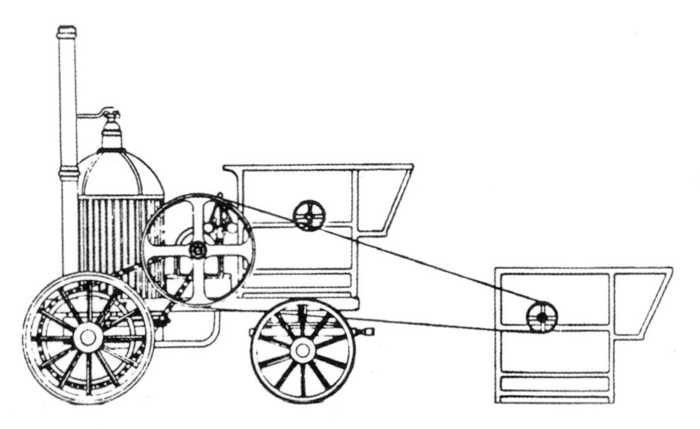

Ransomes & May made a self-propelled version of their portable engine with a chain drive to the front wheels in 1842.

A Ransomes & May 5nhp portable engine was shown at the Great Exhibition in 1851.

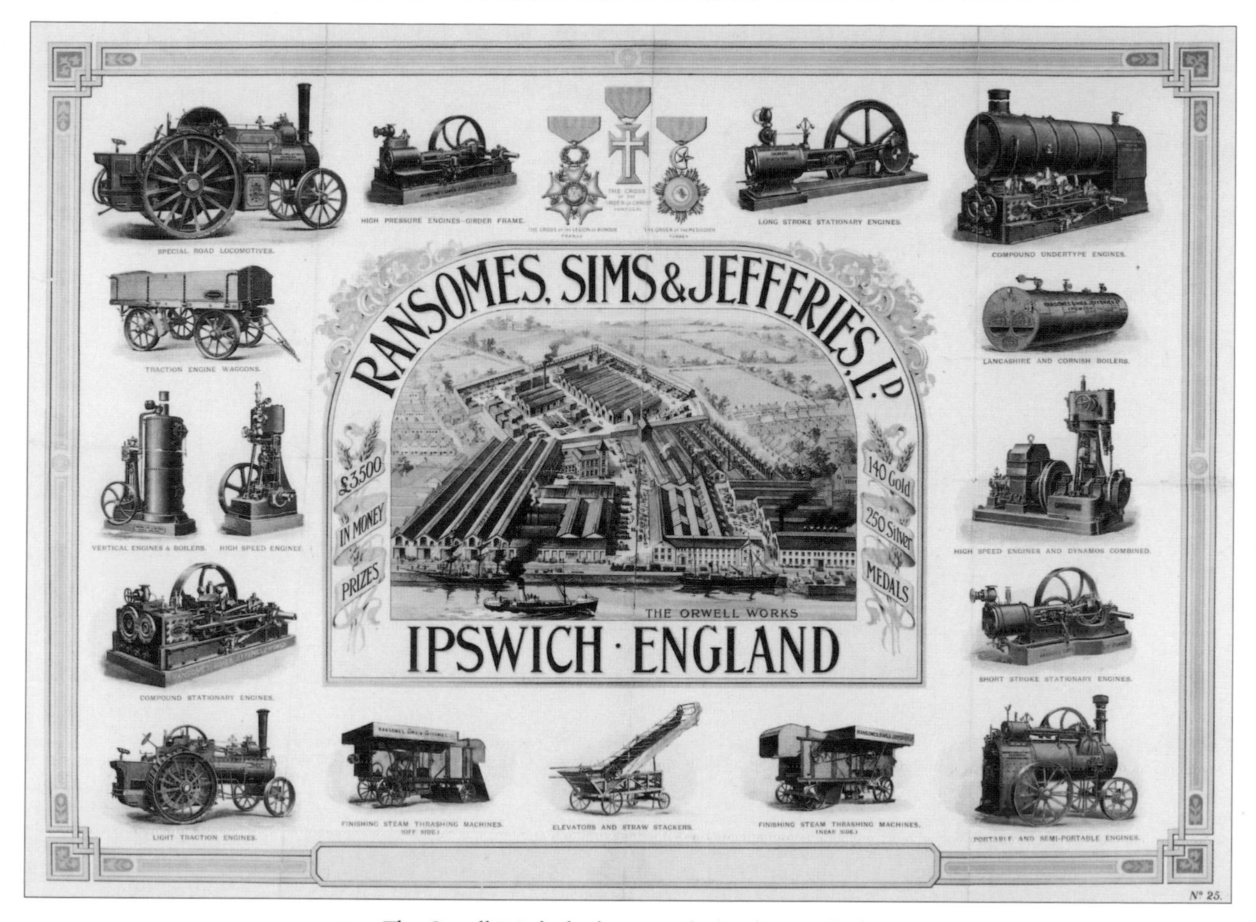

The Orwell Works had a great deal to be proud of.

NOMINAL HORSE POWER

Steam engine builders used nominal horse power, developed by the Royal Agricultural Society in the 1840s, to enable farmers to compare the power of a steam engine with that of a horse. The power of internal combustion engines is calculated in brake horse power. One nominal horse power [nhp] is roughly equivalent to 7 brake horse power [bhp], so a 5nhp Ransomes & May portable steam engine made in 1851 had approximately the same power output as a modern 35bhp farm tractor.

As previously mentioned, John Fowler, who was already selling his own designs of horse-drawn land-drainage equipment in the early 1850s, co-operated with Ransomes in designing a cable-drawn mole drainer built for Fowler at the Orwell Works. On holiday at Brighton in 1852, John Fowler by chance met William Worby, who was implement expert and factory manager for Ransomes & Sims, and they went on to discuss problems of the design and draft of ploughs and cultivating machinery. The meeting eventually resulted in Ransomes building the first of a number of four-furrow balance ploughs to John Fowler's design as well as single-cable Fowler ploughing engines in 1856. This arrangement continued until 1862 after Fowler had established his own steam plough works at Leeds. Ransomes decided against any further development of this type of ploughing engine and concentrated on building traction engines for thrashing, haulage and other farm work.

The 1860s were a period of rapid growth in portable and traction engine production at Ipswich. A 25 ton Ransomes traction engine, introduced in 1862, inherited various features from portable engines including a flywheel at the front end of the boiler and the cylinder mounted above the firebox. The engine had large-diameter, gear-driven rear wheels, and a man

More than two thousand Ransomes portable straw burning engines were in use in all parts of the world in the late 1880s. The smallest 4nhp single-cylinder engine cost £170 and the largest 20nhp double cylinder model was £470. A charge of £22 10s was made for packing the 20nhp portable for shipping overseas.

standing on a platform at the back steered it by means of a system of shafts and chains to the front axle.

Stationary steam engines for factories, cotton mills and coal mines were also built at the Orwell Works. These included single-cylinder 1½ to 10nhp vertical engines with an upright boiler that were mounted on a four-wheel trolley or a solid cast-iron base, as well as one- or two-cylinder 4 to 50nhp horizontal engines on a heavy cast-iron frame. Horizontal engines developing up to 160nhp were built at the Orwell Works in the early 1900s.

Steam power had become an important part of the Ransomes, Sims & Head export business in the 1870s and many of their engines were modified to burn wood, peat, sawdust, straw and other fuels.

An involvement with India was initiated by Lieut. Col. R.E. Crompton of the Indian Army, a leading engineer and an authority on military vehicles, who decided that, because the development of the Indian railway system was slow, a steam-powered road train should be developed for use on the new road being constructed between Delhi and the Punjab. British engineer Robert Thompson had already designed a road-going tricycle-wheeled steam engine and Crompton, being aware of Thompson's steamer, engaged Ransomes to build the Chenab road steamer.

The engine was made at the Orwell Works in 1870 under the supervision of John Head and was tried out on the then Ipswich racecourse. Unfortunately the boiler overheated and hot sparks from the chimney stack set fire to the grandstand. After modifications the road steamer went on a successful test run from Ipswich to Edinburgh and back at speeds in excess of 20mph. In 1871 the Indian Government gave Ransomes an order for four of these tricycle-wheeled road steamers on solid rubber tyres. The wood-burning engines were used to haul large trailers carrying up to 65 passengers or 40 tons of freight between various towns in India at speeds of 15 to 20mph. By the end of the decade, however, the road steamers had been replaced with railway trains on the vastly improved Indian railway system.

Ransomes, Sims & Head were making 6 and 8nhp agricultural locomotive engines in the 1870s. Both were 'arranged to drive at two different speeds, viz. about 1½ and 3 miles per hour', and could pull gross loads of 8 to 10 and 12 to 15 tons respectively.

John Head was also involved in the design of a portable straw-burning steam engine. It was the star attraction at the 1873 Vienna Exhibition and was awarded a gold medal at the Paris International Exhibition in 1878. Several hundred of these engines were sold with thrashing sets for use in the vast grain-growing areas of Argentina and Russia where eight to ten sheaves of straw provided sufficient fuel to thrash a hundred sheaves of wheat.

THE TRACTION ENGINE DESIGNER

The career of William Fletcher illustrates how traction engine designers and engineers moved from company to company, developing their ideas from this varied experience.

William Fletcher was born in 1848 and apprenticed to steam engine manufacturer Marshall's of Gainsborough in 1863. Seven years later he moved to Wallis & Steevens where he designed the first traction engine built by that company at Basingstoke. Following employment with Charles Burrell at Thetford for two years from 1878 and then back at Marshall's, this time as their chief draughtsman, he was appointed Ransomes, Sims & Jefferies' traction engine designer in 1888. Fletcher made various modifications to existing engines before, in 1892, he introduced a completely new model that became a standard Ransomes' design until they gave up production of traction engines in 1934.

Fletcher, who also wrote books and many magazine articles on the history and design of steam engines, left Ransomes 1897 to become traction engine designer for Clayton & Shuttleworth at Lincoln. He took a similar post with Davey, Paxman in 1906 and remained with the Colchester engine builder until his death in 1916.

Improved portable steam engines with the cylinder in front of the boiler, twin flywheels above the firebox and an automatic governor had come into use by the early 1880s. Power output ranged from 3 to 14nhp and Ransomes' portables, with a dynamo mounted on a frame in front of the chimney, were used to generate power for the new electric lights gradually replacing gas lamps in the streets of London.

Early steam engines had a single cylinder but Ransomes and their competitors were building more powerful compound engines in the 1880s. Originally introduced in 1879 by Richard Garrett & Sons at Leiston in Suffolk, the compound engine had two cylinders that used the steam twice. Steam at boiler pressure was first supplied to the smaller cylinder and then, having done its work there, was transferred at lower pressure to the larger cylinder before it was released to the atmosphere.

Stationary steam engines for driving small thrashing machines and other barn equipment appeared in the early 1850s, and by the mid-1880s Ransomes were making nine sizes of stationary compound engine from 8 to 40nhp that cost between £160 and £500. The more powerful engines, often with a locomotive-type boiler, were used for driving pumps, operating hoists and winding gear, and for other industrial applications.

Ransomes also built showmans engines for hauling fairground equipment from town to town and generating electricity for the fairground lights, roundabouts and other rides. The highly decorated showmans engine had large-diameter wheels, an ornate canopy and a dynamo. Less ornate models built at Ipswich included steam-powered road tractors used for general haulage work and steam wagons, and like other Ransomes engines they were exported to many parts of the world.

The 6 and 8nhp Ransomes agricultural locomotive engines weighing 7 and 9 tons respectively were being made in the early 1880s. The smaller engine hauled up to 10 tons and the larger model had a payload of 12 to 15 tons. Both had road speeds from 1½ to 3mph and the facility to attach a wire-rope winding drum to the main axle to move a thrashing drum while the engine was stationary. The catalogue for 1886 noted that the winding drum was an extremely useful addition to a traction engine as 'all those who are accustomed to them will see at once'. It was also explained that these engines were 'intended for all purposes to which steam could be applied as motive power for farm work' and were 'very simple and easy to manage, and do not require a skilled mechanic to work them.'

The Ransomes Colonial traction engine [RCT] designed for road haulage, direct traction ploughing and thrashing, was introduced in the mid-1880s. A large firebox, which could be adapted to burn wood or straw, was a distinguishing feature of Ransomes' 15 to 60nhp Colonial engines with a single or compound engine and a friction clutch in the flywheel.

A few crane engines were also built during this period, No. 31066 (Page 12) being used to load railway wagons at the Orwell Works. Because the weight of the front-mounted crane made the engines difficult to steer, a mechanical power-steering system driven from the crankshaft was devised for them.

The Heavy Motor Car Order of 1903 allowed light road engines to travel at 5 mph on the

Ransomes' 4nhp-steam tractor No. 23266 was built in 1909 and exhibited at the 1910 Smithfield Show. It was used for haulage work and shunting railway wagons at the Orwell Works between 1912 and 1942 when a Fordson tractor replaced it. Following a period of retirement No. 23266 was restored and named the Back'us Boy by Mr Humphrey Dawson at the 1960 Suffolk Show. For many years it was owned by Ransomes' employee Doug Cotton and it is still to be seen at steam rallies in the UK and in Europe.

The Mendip Lady.

public highway and Ransomes took the opportunity to introduce a new 5 ton steam motor tractor with a compound engine. They also produced a 5 ton overtype steam wagon with the engine mounted in the cab above the boiler, and drum brakes on the rear axle. They also built undertype steam wagons which had the boiler in the cab but the engine was mounted under the load platform. Various body types were used including a fixed load platform and two- or three-way tippers with a mechanical or hydraulic tipping mechanism.

The internal combustion engine was already replacing steam power in 1919 when Ransomes formed an association with Ruston & Hornsby of Lincoln in an attempt to standardise production of steam engines and other machines and enable both companies to benefit from mass production and achieve economies in design and marketing. Steam wagons, including a new overtype model exhibited at the 1920 Smithfield Show, resulted from the association but, with petrol-engined lorries coming into fashion, the days of steam were numbered and the last Ransomes steam wagon was built in 1927 and exported to Australia.

Road Roller No. 19609 was the only Ransomes-designed roller built at Ipswich. It was a convertible engine built in 1907 and based on a 7nhp traction engine. The front roll and saddle could be substituted with wheels and steering and a pair of traction wheels replaced the rear rolls. Although this was the only roller designed by Ransomes, a number of Ruston-design rollers were built at the Orwell Works in the late 1930s for Aveling-Barford at Grantham.

The last traction engine sold on the home market was built in 1934. It was driven from the Orwell Works by Fred Dyer and exhibited at the Royal Show held in the same year at Ipswich. A few Ruston & Hornsby design portable steam engines were made by Ransomes at Ipswich during the latter part of the 1930s. One of them, built for a Dutch customer in 1939, was not delivered until well after the end of hostilities in 1948. Ransomes' association with Ruston & Hornsby ended in 1940 when the Lincoln company made a similar arrangement with Davey, Paxman & Co. at Colchester.

Ransomes sold the steam engine side of their business to Robey & Co. Ltd at Lincoln in 1956. Established by Robert Robey in 1854, this company made their first portable engine in 1885 and traction engines from 1891. Robeys built a few traction engines for export in the late 1940s and they made about thirty Ransomes portable engines. The last four portables, built by Robeys in the early sixties, spent their working life in the Sudan.

Oil Engines

A 12bhp horizontal semi-diesel internal combustion engine made in 1914 was the first of a series of six Ransomes oil engines in the 10 to 30hp bracket.

The four-stroke hot-bulb engines ran on crude oil. They were started by using a blow lamp on a bulbous extension to the cylinder head to create sufficient heat inside the cylinder to ignite the low-grade fuel drawn into the engine when it was cranked by hand. Stationary and portable versions were made, the portable engines being mounted on a four-wheeled carriage with shafts for a horse or oxen. The 10 and 12hp stationary engines had two 4ft diameter flywheels, and the 15, 20, 25 and 30hp models had a single flywheel with diameters varying between 5 and 6ft. They were tank-cooled and, depending on horsepower, had two, three or four water tanks. Ransomes' stationary oil engines were very heavy: the 10hp model weighed 34cwt and the 30hp engine tipped the scales at 4 tons.

Ransomes' portable oil engines had two flywheels and

were carried on a steel girder chassis. Fuel and cooling water tanks were mounted on the chassis and a pump circulated the water through the cylinder jacket and returned it to the cooling tank. The 10 hp model, with two 3ft 9in diameter flywheels, weighed 50cwt and the largest 30hp engine with 4ft 8in diameter flywheels weighed a massive 95cwt.

Large numbers of these oil engines were exported and customers were advised to include with their order a full set of spares including big end and main bearings, piston rings, valves and glass lubricators.

The 3½/4hp and 7/8hp two-stroke vertical Wizard water-cooled stationary engines running on paraffin appeared in 1921. During their six-year production run some five hundred were made about half of which were exported, mainly to Australia and Argentina. The Wizard was sold either as a thermo-syphon-cooled stationary engine on a cast-iron base or as a radiator-cooled portable on wooden skids or a trolley. The cooling water tank was about 4ft high on the 3½/4hp engine and 6ft 6in high by 3ft in diameter on the 6/7hp version. The Wizard differed from other engines of the day in that instead of the usual hot bulb or magneto it had the Brons starting system developed in Holland. The engine was run on paraffin and started on it. New owners were warned that under no circumstances should they attempt to start it on petrol.

About 500 Wizard stationary engines were made between 1921 and 1927.

The Wizard was a high-compression engine with a compression pressure of about 450psi. No external heat was required to start the engine and instead of paraffin being introduced into the cylinder at the end of the compression stroke, the fuel was drawn into a porous cup during the suction phase. When the engine was near its maximum compression pressure, fuel was forced through the pores in the cup and mixed with the high-temperature air inside the cylinder.

It was started by de-compressing the engine and slowly cranking it for a few turns to draw fuel into the cylinder. Then the cranking speed was increased for another two or three revolutions before releasing the de-compressor lever. If all was well the engine would fire and run. It was also possible to start the Wizard backwards and when this happened it ran on the oil in the sump instead of paraffin. In this situation the machine was likely to disintegrate if it was not brought to a rapid stop and more than one farmhand had to crawl across a barn floor to avoid being hit by an orbiting crank handle while attempting to stop a runaway Wizard.

Huge numbers of small four-stroke lawn mower engines were made at the Orwell Works in the 1930s, and the 600cc four-stroke engine for the MG 5 launched in 1948 was another Ransomes product. Five years later, the company was making V12 diesel engines for Royal Navy minesweepers in new production facilities at Nacton Road. The engines were designed by Paxmans of Colchester who in conjunction with Ransomes and Rustons of Lincoln built several hundred of these engines between 1953 and 1956. Aluminium was used for the engine castings and other parts were made from special non-magnetic iron and steel - an important consideration when building wooden-hulled minesweepers used on missions to search for and destroy magnetic mines.

Tractors

The first tractor with an internal combustion engine was probably made in America in 1885 and the Hornsby-Ackroyd of 1897 was the first British-built oil-engined tractor. Six years later, in 1903, James Edward Ransome took personal charge of the design of a 20hp tractor with a four-cylinder petrol engine with chain-driven rear wheels. It had a friction clutch, three forward gears and reverse, hand brake, foot brake and an expected price of £450. Local farmers were invited to a working demonstration of the tractor with a three-furrow chain-pull riding plough but horses and steam engines were still the acknowledged source of farm power and little more was heard of James Ransome's tractor.

A second assault on the tractor market came some ten years later. This time the tractor had a 12in diameter cylinder with a 14in stroke and developed about 35hp. It was about 18ft long, 7ft wide, 11ft to the top of its exhaust pipe and weighed about 10 tons. With such oversize dimensions this Ransomes tractor was also destined to fail and several years passed by before farmers accepted the tractor as a serious alternative to the horse.

Ransomes' next experiment with petrol-engined tractors came in the early 1930s when they built and tested a prototype pedestrian-controlled garden tractor with rubber-jointed tracks made by Roadless Traction Ltd at Hounslow in Middlesex. This project was dropped, and they turned their attention to building and testing an experimental ride-on garden tractor with Roadless rubber-jointed tracks.

The Ransomes Motor Garden Cultivator, demonstrated for the first time on a farm near Evesham in Worcestershire on 29th April 1936, was the first riding tractor designed specifically for the smallholder, market gardener and fruit farmer. The *Roadless News,* issued by Roadless Traction Ltd, reported that the Motor Garden Cultivator, 'designed to take the place of two horses for market gardening and fruit growing, received a great reception' at the demonstration. The writer commented that the tractor was so easy to control that even on his first run he was able to turn it without difficulty and that the headland was practically non-existent. In 1937 *Roadless News* reported that one large-scale East Anglian farmer was using six Ransomes Motor Garden Cultivators for all inter-row cultivations in his sugar beet crop.

The Ransomes MG [Motor Garden] Cultivator, later known as the MG 2, had a 600cc air-cooled, single-cylinder Sturmey Archer 'T' engine, one forward and one reverse gear and 6in wide Roadless rubber-jointed tracks. The 6hp side-valve engine had a dry sump with a separate tank for the lubricating oil. This was pumped to the bearings and a second pump retrieved the oil and returned it through a filter to the tank. The engine had a Lucas magneto without an impulse coupling, and the starting handle dog was attached to a countershaft connected by a roller chain to the crankshaft. Power was routed through a four-to-one reduction gearbox on the engine output shaft to a centrifugal clutch that automatically engaged the drive to the tracks when the engine speed reached approximately 500rpm. The MG 2 had a top speed of

Ransomes, Sims & Jefferies demonstrated a 20hp petrol-engined tractor on nearby Rushmere Heath in 1903.

Sales literature suggested the MG 2 was a 'high class baby track-type tractor so simple that a boy could operate it'.

about 2mph in both directions and a single lever was used to select forward, neutral or reverse. The transmission consisted of an inward-facing pair of crown wheels and a driving pinion on the clutch output shaft. The lever was used to engage the pinion with one of the crown wheels to select forward or reverse. It was steered with two lever-operated band brakes on the drive shafts to the tracks and both shafts were always under some degree of power to eliminate slewing or sliding when the tractor changed direction. The centrifugal clutch served as a safety overload mechanism so that when the tractor was seriously overloaded, engine speed fell away and the clutch disengaged the drive. The tracks had a ground pressure of about 4psi and the track width could be adjusted with different-sized spacing blocks between the track mounting points and the chassis, giving three settings between 2ft 4in and 2ft 10in.

An improved Sturmey Archer 'TB' engine replaced the earlier 'T' power unit in 1938. It had a Wico magneto with an impulse coupling and a belt-driven cooling fan. Gears replaced the chain drive from the starter dog to the crankshaft.

Sales literature explained that in an eight-hour day the MG 2 could

The instruction book for the MG 5 explained that the tractor was designed to do 'two-horse work at two-horse speed', but to keep up with the advance in farm mechanisation its speed had been increased from 2 to 2½ mph.

Ransomes offered a full range of implements for the MG 5.

plough an acre with a single-furrow plough or cultivate five acres with a seven-tine mounted cultivator. Complete with a hand-lift tool frame and swinging drawbar, the MG 2 cost £135 in 1936, an optional 400rpm power take-off shaft adding £1 10s to the price. A considerable number of MG 2s were used in French vineyards and Ransomes were astute enough to provide both imperial and metric dimensions and capacities in their sales literature. About three thousand MG 2 tractors were produced between 1936 and 1948 and, although materials and labour were in short supply, approximately 1,200 of these were made during World War II.

Implements for the MG 2 included the TS 25 one- and two-furrow hand-lift ploughs, disc harrows and a mounted tool frame with cultivator tines, hoe blades and ridging bodies. The TS 42 self-lift single-furrow plough with an optional subsoiling tine was added a year or so later.

The MG 5, designed to do 'two-horse work at two-horse speed', superseded the MG 2 in 1948. The most obvious visual differences were the fuel tank located under the seat and a cowling over the 600cc Ransomes air-cooled petrol engine. The dry-sump lubrication system was retained along with the four-to-one reduction unit and the single forward and reverse gearing. A vaporising-oil conversion kit was available at extra cost and power take-off speed was increased to 700rpm. The drawbar and hand-lift toolbar used on the MG 2 were retained and an optional hydraulic lift unit made by R.J.Neville in

The MG 6 market garden cultivator with a petrol or tvo engine and hand-lift toolbar cost £305 when introduced at the 1953 Smithfield Show. A power take-off shaft and hydraulic linkage added £52 to the bill.

Shunting railway wagons was one of the tasks performed by the Industrial Wheeled [ITW] version of the MG 6.

Australia added an extra £89 15s to the basic price of £250. A belt-driven hydraulic pump supplied oil from a separate reservoir to an external ram on the rear toolbar and a lever provided fingertip control of the toolbar or a small front dozer blade. Publicity material explained that the MG 5 was the complete answer to the mechanisation of holdings of up to twenty-five acres and was so simple that a boy could operate it with ease.

A wider range of implements for the MG 5 included an improved TS 42A single-furrow plough, HR4 disc harrows, C29 toolbar with cultivator tines, hoe blades, ridging bodies and a potato-raising plough. Orchard-spraying equipment made by Coopers of Wisbech and a small trailer were among a list of approved implements made for the tractor by other companies.

The MG 6 made its debut at the 1953 Smithfield Show. The 600cc Ransomes side-valve petrol or paraffin engine and centrifugal clutch were retained but top speeds of 1⅛, 2¼ and 4mph were provided by a new three-forward and three-reverse gearbox. An optional 8½hp Drayton two-stroke, overhead-valve air-cooled diesel engine became available for the MG 6 in 1956. Hydraulic three-point linkage, parking brake and 700 rpm power take-off, now with an optional belt pulley, were still listed as optional equipment. Other extras included parking brake, a belt pulley, a pair of front steadying wheels, detachable rubber track pads and 8, 10 or 12in wide hardwood track blocks for boggy or swampy land. The front steadying wheels could be used to stop the front of the tractor 'digging in' on steep downhill slopes or undulating land, and detachable rubber pads were recommended when using the MG 6 on hard roads or inside buildings where the tracks might damage the surface. Sales literature also suggested that the rubber pads would reduce track wear and improve adhesion in the field.

The TS 42A trailed plough was still in

This version of the ITC industrial crawler based on the MG 6 with a 15cu ft capacity dump hopper was developed for the Metropolitan Water Board for the maintenance of their sand filter bed.

A moulded fibreglass bonnet and track guard extensions were added to the list of optional extras for the MG 40 in 1962.

production and tractors with hydraulic linkage could be used for reversible or one-way ploughing by hitching the right-handed TS 65 and left-handed TS 66 ploughs to a modified three-point linkage. An increased range of toolbars included the C29 hand-lift and C67 hydraulic-lift rear toolbars, the C70 and C71 hydraulic- and manual-lift front toolbars and the C78 vertical toolbar for the rear hydraulic linkage. Other mounted equipment for the MG 6 included a low-volume crop sprayer with a 20ft folding spray bar, steerage hoe, rotary cultivator, hammer mill and saw bench.

There was a choice of petrol or diesel engine for the Industrial Tractor Wheeled [ITW] and Industrial Tractor Crawler [ITC] versions of the MG 6 introduced in 1956. Heavy-duty roller chains on the ITW wheeled model transmitted drive from the standard crawler transmission to the front wheels on the skid-steered model used for haulage work and shunting railway wagons. Other equipment included a small dozer blade and rotary sweeping brush.

The ITC tracked version had the same track layout as the standard MG 6, and optional rubber blocks could be bolted to the track plates when working on hard surfaces. A trailer or implement could be towed from the rear drawbar and, with the optional hydraulic linkage, the ITC was used with a front-mounted dozer blade. Alternatively, by modifying the steering levers and seating the driver with his back to the engine, the ITC could be used with a digger loader or a dump hopper mounted on what would normally be the back end of the machine. As the tractor had three

This one-off special with a steering wheel and based on an MG 40 was used to tow barges on a London canal.

forward and three reverse gears with the same ratios, it was equally simple to drive the ITC in either direction. When Ransomes discontinued the ITC and ITW versions of the MG crawler, a limited number of tracked WR4 dumpers and WR8 loaders built on the MG 6 and MG 40 were made by Whitlocks of Great Yeldham in Essex. The MG 40, with the choice of a petrol, paraffin or diesel engine, superseded the MG 6 in 1960. The three engines had wet-sump lubrication with an oil pump and filter and were started with a crank handle. An Amal carburettor and Wico magneto were used on the 8hp four-stroke side-valve petrol and petrol/paraffin engines. An ignition wick was used to cold-start the alternative 10hp two-stroke overhead-valve diesel engine which had a fuel consumption of about 3 pints an hour.

The Roadless Traction rubber-jointed tracks were strengthened on the MG 40; the centrifugal clutch, three-forward and three-reverse gearbox and differential spur-gear reduction units were taken from the MG 6. Improvements in 1962 included new steel track guards. Needle roller bearings were used in the track rollers and idler wheel hubs, and later models had a Sachs diesel engine. The optional hydraulic linkage added £82 2s to the basic price of £500 for the Ransomes MG 40 when it went out of production in 1966.

About 15,000 MG crawlers including 3,000 MG 2s, 5,000 MG 5s and MG 6s and about 2,000 MG 40s were made during a thirty-year production run. Whereas the other models were originally manufactured in the 'E' Department at the Orwell Works and subsequently at Long Street, the MG 40 was made in the lawn mower works. Until 1946 the MG 2 tractors were painted Orwell blue. In that year the colour was changed to the darker Nacton blue and this shade was also used for the MG 5 and MG 6. The MG 40 was once again the exception, painted Ransomes lawn mower green.

MG Serial numbers are listed on page 186.

Compact Tractors

Ransomes returned to the tractor market in 1997 with the introduction of a range of five Japanese-built four-wheel-drive compact tractors. The CT 318, CT 320 and CT 325 had 18, 20 and 25hp three-cylinder diesel

Ransomes introduced a new range of compact tractors in 1997.

engines with manual or hydrostatic transmissions. The 33hp three-cylinder CT 333 HST had a three-speed hydrostatic transmission with cruise control and an independent hydraulic system to raise and lower its mid-mounted mower deck. The 38 and 45hp four-cylinder diesel-engined CT 435 and CT 445 were equipped with shuttle transmissions, mid and rear power take-off shafts, power steering and quiet cabs.

The blades on the Vibro Hoe could be set to work in row crops spaced between six and eighteen inches apart.

Vibro Hoe

Ransomes' sales literature in the mid-1950s explained that the unique principle of reciprocating hoe blades made the self-propelled Vibro Hoe the most efficient hoe and cultivator ever offered to smallholders and horticulturists, and furthermore that it would 'banish the tiresome back-aching drudgery associated with manual hoeing.'

The prototype version of the Vibro Hoe had two wheels but to keep it as narrow as possible the production model ran on a single wheel with a solid rubber tyre. Initially produced in 1954 at the Lorant works at Watford (see page 77), the Vibro Hoe had a two-stroke Villiers engine with a centrifugal clutch to engage drive to the land wheel and the hoeing mechanism. A system of cranks was used to move one blade on its forward stroke while the other moved rearward. This gave the hoe blades a walking action as the hoe progressed between two rows of plants. Short lengths of bicycle-type chain attached to the blades stirred the soil and helped to expose the weed roots. The Vibro Hoe could also be used with cultivator tines and their vibrating action was claimed as an advantage. The throttle had to be fully open when starting the engine, making it necessary to disengage the drive to the wheel and hoe blades with a dog clutch before using the starter rope.

The design of the Vibro Hoe was mechanically suspect and all machines sold were recalled on a couple of occasions for modification at the factory. Production was transferred from Watford to Ipswich in 1956 and the last Vibro Hoe was made at the Orwell Works in 1958.

Chapter 6

Lawn Mowers and Turf Equipment

Edwin Budding of Stroud in Gloucestershire is credited with inventing the cylinder lawn mower. A similar mechanism with spiral cutting knives on a cylinder rotating against a fixed blade was used to shear the nap from cloth in textile factories and Budding adapted this mechanism for his mowing machine. Edwin Budding's partner John Ferabee, who owned the Phoenix Foundry in Stroud, patented the design on 5th October 1830 and reserved the right to make Budding's mower himself or to license other companies to manufacture it. J. R. & A. Ransome were the first company to obtain a licence from John Ferabee and the first Budding's patent mower was made at Ipswich in 1832. An annual production of between seventy and eighty machines continued for the next twenty years.

Ransomes made the first Budding's patent lawnmower in 1832.

Budding wrote that 'country gentlemen would find using his machine an amusing, useful and healthful exercise' and a Ransomes catalogue suggested that 'persons unpractised in the art of mowing with a scythe may now cut the grass on lawns, pleasure grounds and bowling greens with ease.' The following instructions were provided on how to use the machine: 'Place it so as to bear the whole weight on its cast-iron drum and on the wood roller, then take hold of the handles, as in driving a barrow, and with the right thumb, pressing the end of the bent lever towards the right side of the frame, push the machine steadily along the greensward, without lifting the handles, but rather exerting a moderate pressure downwards, not so much however as to lift the wood roller off the ground, which would raise the cutters above the grass. The grass is best cut when dry and not too long.'

Other instructions informed users that the length of grass left on the turf depended on the distance the wood roller was set below the frame, and that the ratchet drive to the cylinder was disengaged by pulling the mower backwards while operating a small lever with the right thumb. Recommendations included the occasional application of sweet oil. When the revolving blades were 'worn by long use', they should be lowered with two adjusting screws placed at each end of the knife roller until they just touched the bottom plate.

THE FIRST MOWER IN THE WORLD

(From Ransomes' catalogue of Motor Mowers, March 1933)

In 1832 a new industry was created when Ransomes made the first lawn mower and in the fullness of time put an end to the manual labour of lawn cutting. Right from the start this invention was a success, its development soon brought home to the public the potentialities of the well kept lawn and gradually enabled ever increasing numbers to enjoy this former luxury.

By 1902 Ransomes had developed and marketed their first Motor Mower and some years later they introduced electrically driven models - again the first in the world.

During their 100 years of accumulated experience Ransomes have studied the requirements of all users and designed and perfected mowing machines for all purposes including Private Lawns; Fairways and Greens of Golf Courses; Sports Grounds; Recreation Grounds; Parks, etc and to-day they offer the largest and most comprehensive range.

The fact that spares for machines sixty years old are not infrequently ordered proves the great durability of Ransomes Mowers.

Ransomes have supplied Motor Mowers to HM King George V and many other Monarchs, also to HRH The Duke of York and to the nobility, Municipal Authorities and many of the famous Sports Organisations.

And so it is that all products of the House of Ransomes, Sims & Jefferies can justly be claimed to be built by Pioneers upon unrivalled experience and relied on for unrivalled service and satisfaction.

Licences to manufacture Edwin Budding's mower were also granted to Thomas Green of Leeds in 1835 and to Alexander Shanks at Arbroath in 1839, who patented a horse-drawn version of the mower in 1842.

Ransomes & May's 1851 catalogue, published in time for the Great Exhibition at Hyde Park in London, included the 16in Budding's grass cutter priced at £6 5s. The 19in machine was £6 15s and the 22in cut mower cost £7 5s. It was noted that this 'unique and valuable adjunct to the pleasure ground and garden of the nobleman or gentleman, has now stood the test of twenty years experience and continues in as high a repute as ever.'

A number of improvements to the 21in cut Budding mower were introduced over the years and by 1852 Ransomes & Sims had built about 1,500 Budding pattern mowers. Ransomes stopped making lawn mowers in 1858 but they were still made by Thomas Green and Alexander Shanks who supplied Ransomes with machines for resale to their customers.

Lawn mower production resumed at Ipswich in 1861 with an improved version of the Budding mower, followed in 1867 by chain- and gear-driven Automaton cylinder mowers. A descendant of the Budding machine, the Automatons were the first lawn mowers designed by Ransomes & Sims. Sizes varied from an 8in mower for small lawns that cost £2 15s to a 48in cut horse-drawn machine priced at £32. A catalogue at the time listed the Automatons according to their power requirement. A lady or boy could work the 10in mower, a lad could push the 12in model and the most popular 16in cut Automaton, which cost £6 10s, required a man to push it and a boy in front to pull it with a rope.

Automaton mowers dominated Ransomes' catalogues for thirty years. They were lighter and shorter than other machines on the market at the time, and they set the standard for hand roller mowers used by gardeners in many parts of the world for the next hundred years.

Ransomes, Sims & Head's Horse Power cylinder lawn mowers for large lawns, tennis courts and golf

Ransomes Improved Patent Automaton mowers were made for thirty-six years.

courses were made from 1870 until the late 1920s when they were superseded by Ransomes, Sims & Jefferies' gang mowers and ride-on motor mowers. A pony was used to pull the 26 and 30in cut machines but a horse was needed for the wider Horse Power mowers that had cutting widths in 6in steps from 30 to 48in. The operator walked behind and steered the mower with a patented spring-loaded handle said to remove 'all disagreeable vibration from the machine and the hands of the user'. It was common practice for the horse to wear a set of leather boots to prevent undue damage to the turf. The 48in Horse Power mower with a side-discharge grass box cost £32 and a set of boots for the horse was £1 5s.

The opening of a new lawn mower factory in Waterworks Street at Ipswich in 1876, where mowers were made for the next sixty years, coincided with the introduction of the Globe mower for cutting long grass Unlike other Ransomes mowers, the Globe (later renamed the World), had a three-knife cylinder as well as two side wheels which replaced the usual wooden front roller. The Little Gem and the Paris side-wheel mowers appeared a few years later. As its name suggests, the Paris mower was specifically made for French gardeners.

In response to the large numbers of American side-wheel mowers arriving in the country in the 1880s, Ransomes' 1883 sales leaflet explained the benefits of buying one of their own machines. It pointed out that the cutting cylinder on the World mower was simple to adjust to the bottom blade. Furthermore, the steel knives were less likely to break than the cast-iron knives, or thin steel knives bolted to cast-iron barrels, on similar lawn mowers imported from America. The World mower was suitable for cutting wet, dry, long or short grass without clogging and was 'well adapted for getting over a large amount of work with little labour.'

IPSWICH. RANSOMES, SIMS & JEFFERIES, LIMITED. LONDON.

LAWN MOWERS.

Awarded a SILVER MEDAL at the Inventions Exhibition, 1885, and confidently recommended as the best Lawn Mower in the world.

THE "NEW AUTOMATON."

These Machines, in point of design, mechanical construction, materials, workmanship, finish and durability, are without a rival. The motion is imparted to the knives by accurate machine-made GEARING, which works smoothly and with little noise. It is completely covered, and by the very free motion of the knives they entirely clear themselves from all cut grass. The draught has been reduced to the smallest amount compatible with the durability of the Machine made in the following sizes :—

8-in.	For small Lawns.	16-in.	For Man and Boy.
10-in.	,, Lady or Boy.	18-in.	,, ,, ,,
12-in.	,, Lad.	20-in.	,, Two Men.
14-in.	,, Man.	22-in.	,, Donkey.
		24-in.	,, Small Pony.

THE "CHAIN AUTOMATON."

These Machines are the same in general construction as the "NEW AUTOMATONS" described above, but are driven by a CHAIN instead of gearing.

10-in.	For a Lady or Boy.	14-in.	For a Man.
12-in.	,, Lad.	16-in.	,, Man and Boy.

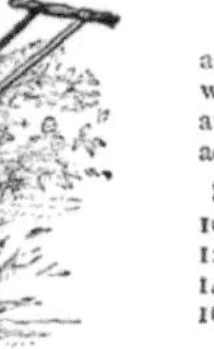

THE "WORLD."

These Mowers are intended for cutting long grass and doing rough work, and are adapted to compete with the Lawn Mowers imported from America, but are better fitted, more durable, and have more perfect adjustment.

8-in.	For Small Lawns.	18-in.	For a Man and Boy.
10-in.	,, a Lady or Boy.	20-in.	,, ,, ,,
12-in.	,, ,, ,,	22-in.	,, Two Men.
14-in.	,, Man.	24-in.	,, ,,
16-in.	,, ,,		

Grass Boxes placed in front of the Machine, extra.

THE "NEW PARIS."

Light, simple, and low-priced Machines for small gardens and amateurs' use. They will cut long, short, wet, or dry grass, and, like all R. S. & J.'s Lawn Mowers, are driven from both sides, which specially adapts them for cutting round borders, trees and shrubs.

SIZES :—6, 8, 10, 12, 14, and 16 inches.

"PONY AND HORSE-POWER."

Suitable for Large Lawns and Pleasure Grounds, Cricket Grounds, Arboretums, Lawn Tennis Clubs, etc.

They can be fitted with patent apparatus for clearing the grass box from the handles without stopping the horse, and with seat for the driver to ride, at a slight additional cost.

Pony Machines, 26-in. and 30 inches wide.
Horse Power, 30, 36, 42, and 48 inches wide.

Special circulars of Lawn Mowers on application.

A page from Ransomes, Sims & Jefferies' catalogue for 1886.

Gear- and chain-drive New Automaton mowers with nine cutting widths replaced the original Automaton in 1885. Gear- and chain-drive models were the same price. The smallest 9in model for small lawns and edgings cost £2 15s. Two men or a pony were needed to for the 20, 22 and 24in mowers and the biggest 24in cut New Automaton with a pony whippletree was £10 15s.

Improved Patent Chain Automaton and Patent Gear Automaton mowers with ribbed driving rollers

Cutting the lawns at Balmoral with a Ransomes Horse Power mower in the early 1900s.

replaced the New Automatons in 1894. The catalogue described its patent single-screw adjustment for setting the cutting parts as 'the simplest arrangement that has ever been brought out'. The blades on the cutting cylinder were made in two parts with both halves angled inwards to direct the clippings centrally into the grass box.

The Lion and Anglo-Paris side-wheel mowers were other long-lived Ransomes machines. The Lion appeared in 1895. More than a hundred thousand had been sold by 1915 and it remained in production well into the 1930s. The Anglo-Paris, introduced for the home market in 1892, was similar to the Paris and the improved Mk II Anglo-Paris with larger side wheels was still being made in 1914

Ransomes, Sims & Jefferies were a market leader in the world of lawn mowers in the mid-1880s and Royal recognition came in 1889 when Queen Victoria bought a Ransomes mower at the Royal Agricultural Show of England held that year at Windsor.

Prices of World Lawn Mowers Without Grass Boxes 1883

8in	For small lawns and edgings	£2 2s 0d
10in	For use by a lady or a boy	£3 3s 0d
12in	Ditto	£4 4s 0d
14in	For use by a man	£5 5s 0d
16in	Ditto	£6 6s 0d
18in	For use by a man and a boy	£7 7s 0d
20in	Ditto	£8 0s 0d
22in	For use by two men	£9 0s 0d
24in	Ditto	£10 0s 0d

Grass boxes extra if required

8in – 5s 0d   10 to 22in – 7s 6d   24in – 10s0d

The company took full advantage of this success and an advertisement at the time listed many important people, including the Prince of Wales, the Duke of Edinburgh, various Earls, Dukes and Peers of the

Realm, who had purchased a Ransomes mower. Other influential users included Harrow School, Oxford University, Kew Gardens, Regents Park and the Exhibition Gardens at Melbourne in Australia. With such an impressive list of customers, the publicity department considered it unnecessary to provide any technical information in their advertisements, relying instead on scenes of fashionable ladies and neatly trimmed lawns.

Although the grass on many golf courses was still cut by hand in the late 1890s, a hundred or so of the more wealthy clubs were using Ransomes mowers. The tees were usually cut with an Anglo-Paris, and a 16in gear-driven New Automaton, used by a man and a boy, was generally preferred for cutting the greens. The fairways were kept in trim with a two-man 24in cut New Automaton or a 48in cut Horse Mower. The improved horse-drawn Ideal cylinder mower for rough grass, introduced in 1905, had its cutting cylinder unit hinged on to the main frame, enabling it to follow the 'inequalities of the ground'. Advertisements pointed out that the Ideal had a comfortable cast-iron seat.

Steam was still the major source of power in the 1890s. A steam-powered, pedestrian-controlled lawn mower weighing 1½ tons was made at Leyland in Lancashire in 1892 and Alexander Shanks at Arbroath built a steam-driven ride-on lawn mower in 1900. However, petrol engines were gaining in popularity and James Edward Ransome, who realised that the future lay with these lighter and more easily operated engines, made the world's first commercial motor mower in 1902. His first customers were Mr Prescott Westcar of Herne Bay, who bought a 42in cut Ransomes Patent 6hp Motor Mower, and Cadburys who bought the second machine for their Bournville sports grounds. Simms made the four-stroke water-cooled petrol engine in Germany but from 1903 it was built under licence at Ipswich and called the Orwell engine. The mower was steered with a handle linked to a small roller under the seat. The grass box, similar to the side-discharge box used on the Horse Power mower, was emptied by pulling a board sideways across the box with a crank handle and endless roller chain.

Within twelve months Ransomes' catalogue listed four different motor mowers with prices ranging from £75 to £150. The 36 and 42in cut mowers for golf courses had a trailing seat, and a 30in cut pedestrian-controlled machine was recommended for cutting grass on 'steep inclines and golf links'. The 24in cut motor mower was described as 'the most useful size for an ordinary garden where there is much grass to cut, and for tennis courts'. King Edward VII requested a demonstration of the 30in motor mower in 1904 and two were purchased for the Buckingham Palace gardens.

More than six hundred motor mowers with Ransomes Orwell petrol engines had been made at Ipswich by 1914 when much of the Orwell Works was turned over to armament production. However, in spite of the country being on a war footing, Ransomes' lawn mower price lists for 1915 included 24in and 30in cut motor mowers for £85 and £130 respectively, while

The grassbox on this 1902 Ransomes motor mower has an endless chain from a board at one end round the crank wheel and then to the opposite side of the box. It was emptied by sweeping the clippings sideways on to the ground.

Ransomes side-wheel mowers with a single six-foot-long handle, known as bank mowers, cost between 2s and 2s 6d more than a hand mower.

£200 was being asked for the 42in cut motor mower with a 10hp twin-cylinder water-cooled engine and trailing seat. Pony and Horse mowers were considerably cheaper with prices ranging from £19 to £42, and the 24in cut Automaton cylinder mower cost £10 15s.

Ransomes' side-wheel mowers for 1915 included the New Empire, Star and Lion. The New Empire, an improved version of the Empire Major introduced in 1903, was a seven-knife cylinder mower with high side wheels in 9, 11, 13 and 15in cutting widths. The Star, with enclosed gearing, was made in 10, 12, 14 or 16in cutting widths. The side-wheel Lion, described as 'affordable for all', cost £1 6s for the 9in cut mower and £1 12s for the 15in model, but a grass box was not included in the price. Ransomes had made more than 100,000 Lion mowers by 1915, but as an agricultural worker's weekly wage was only 18s it is likely that very few cottage gardeners were able to afford the luxury of a new side-wheel mower.

The first professional electric-powered lawn mower was probably made in Australia in 1919, and Ransomes introduced the first British-built mains-electric cylinder mower in 1926. The 14, 16 and 20in cut Electra mowers with a 1½hp electric motor were intended for domestic use and a special 20in model called the Bowlic was produced for cutting bowling greens. The relatively lightweight Bowlic, which did not spill oil or petrol on the turf, was popular with green keepers and remained in production for many years.

The mains electric cable was supported by a swinging arm on a mast attached to the mower. On reaching the edge of the lawn, the operator was meant to use the handle provided to swing the arm across to the other side of the mower and move the cable away from the cutting cylinder while turning the machine for the next run. The 14in model was chain driven from the motor shaft to the cylinder and then to the drive roller; the other sizes included a 30in machine available on special order, which had the same type of enclosed driving gears as were used on motor mowers.

Horse and pony mowers remained in production until the mid-1920s when Ransomes introduced a ride-on pusher attachment for converting pony mowers to petrol power. Attached to the rear of the mower, the pusher had a petrol-engine-driven land roller with a seat on the roller frame. The mower was steered from the seat and the operator could disengage the drive when the grass box was full.

A new era in the maintenance of sports grounds arrived in 1914 when The Worthington Mower Co. at Shawnee in Oklahoma patented a horse-drawn gang mower with three or more side-wheel cylinder mowers ganged or linked together on a frame. Ransomes were granted a licence to manufacture Worthington gang mowers in 1921 and within ten

Three models of the Electra mains electric lawn mower were made in the early 1930s. The 14in cut model had 25 yards of cable; 50 yards were supplied with 16 and 20in cut machines. The 20in model weighed 3cwt, cut about half an acre an hour and used one unit of electricity.

years Ransomes' Triple, Quintuple and Septuple gang mowers with overall cutting widths of 7ft, 11ft 6in and 16ft respectively were cutting sports grounds and parks in many parts of the world. The horse was not always the best means of pulling a gang mower and within a few years agricultural tractors or road vehicles with a suitable towbar were being used. The Nonuple gang mower with nine units and a cutting width of 20ft was added to the range in 1939.

Ransomes triple mowers were made with a tractor drawbar or a sulky hitch for a horse.

The 14 and 16in cut Certes greens hand mowers with eight-knife cylinders were introduced in 1924 to meet a demand from cricket and bowls clubs for a better finish to wickets and bowling greens. Certes mowers with ten-knife cylinders to give an even finer finish on tennis courts and cricket squares appeared in 1932, and with occasional improvements Certes mowers were made for thirty-four years.

The 10, 12 and 14in cut R.S.J. De-Luxe Centenary hand mowers were introduced in 1932 to commemorate one hundred years of lawn mower production since J. R. & A. Ransome made their first Budding pattern machine in 1832. Centenary mowers had dust-proof, self-aligning ball bearings, an enclosed gear drive, a five-knife cutting cylinder and adjustable handles. The14in cut machine with a grass box cost £8 15s.

Ten different models of roller and side-wheel hand mowers were illustrated in Ransomes' 1933 catalogue. They included the Anglia, R.S.J. De-Luxe Centenary and Ajax hand mowers for private gardens, and Certes fine turf mowers. The 12in Ajax, introduced in 1933 with a £4 4s. price tag, remained in production until 1970. Side-wheel mowers included the Leo and Cub for small gardens and the Coronet with a five- or seven-knife cylinder for putting greens and tees. The Kutruf with a large-diameter four-knife cylinder was suitable for cutting rough grass up to 7in high.

Advertising the Leo and Cub side-wheel mowers.

The 15 and 18in cut Certes bowling green mowers were 'of exacting design to ensure an extremely fine finish which is essential to greens laid with Cumberland turf'. The gear-driven ten-

Primarily designed for the Certes mower, Ransomes' grinding-in rest was described as 'a simple contrivance to facilitate the grinding of hand roller mowers when the cylinder knives and bottom blade become dull'. Complete with the hardwood crank handle, the sharpening aid cost 10s 6d in 1933.

knife cutting cylinder ran against a thin, reversible bottom blade at high speed; a single-screw adjuster was used to set the clearance between the cylinder and bottom blade, and another fine-screw adjuster was provided to regulate the height of cut.

In the mid-1930s there was a Ransomes motor mower for every size and type of lawn. Ideal for the smaller lawn was the 14in cut Midget with a 1hp air-cooled two-stroke Villiers engine that cost £22 17s 6d, while for large country house lawns there was the 42in cut motor mower that cost £325 with a five per cent discount for cash. No self-respecting country gentleman would be without the optional trailed seat which added another £6 10s to the bill.

Other mid-1930s Ransomes motor mowers included a 16in model with a Villiers two-stroke engine; 16, 20 and 24in mowers with a four-stroke engine, and a heavy-duty 36in cut machine with a 6hp air-cooled engine. With water-cooled engines, both the 11hp 42in cut combined motor mower and roller as well as the 9hp 36in cut mower were recommended for sports grounds and cricket pitches. They weighed 16 and 22cwt respectively and could cut between 1 and 1½ acres an hour.

The new White City lawn mower works was opened in 1937 to meet the growing demand for Ransomes' grass machinery. Products at the time included the Leo and Ace side-wheel mowers. The Leo, which superseded the Lion in 1932, had easy-running ball bearings, a £2 13s 3d price tag and was advertised as an admirable mower for small lawns. Prices of the Atlas and Anglia roller mowers, recommended for medium-sized lawns, started at £3 8s 6d, and Ransomes' advertisements offered a free demonstration without obligation of these or any other machine anywhere in the country.

New 'Lightweight' 14 and 17in cut mowers with two-stroke Villiers air-cooled engines priced at £18 10s and £32 10s respectively were included in Ransomes' 1939 motor mower catalogue. The 14, 16 and 20in mowers with a two-stroke air-cooled engine, steel side plates and ball bearings were recommended for tennis courts, and a 16 or 20in machine with a four-stroke Sturmey-Archer power unit was the ideal

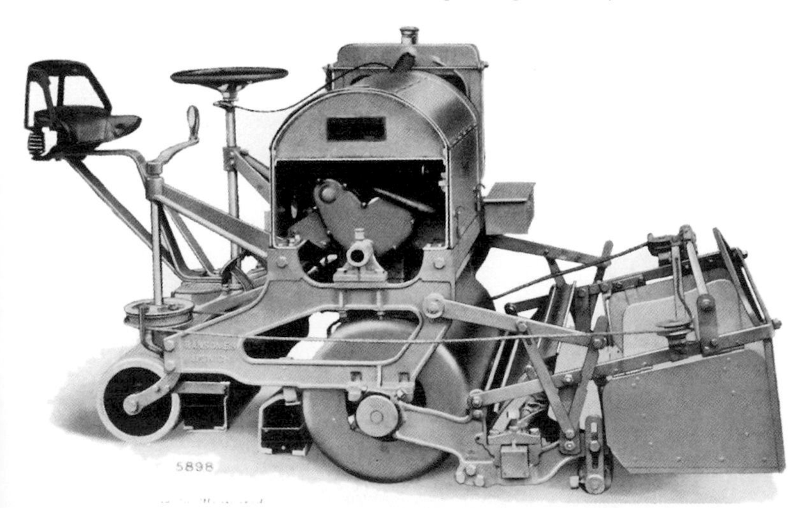

An angled board in the grass box was used to discharge the grass at one side of this 42in cut mower with an 11hp water-cooled engine. The board was pulled across the box with a rope operated by the crank handle on the right of the operator's seat.

The Overgreen greens mower was used on golf greens, croquet lawns and tennis courts.

choice for a large private garden. Professional customers could select a Ransomes 24 or 30in cut four-stroke mower for sports grounds and extensive lawns or an R-type mower for cutting large areas of grass in public parks. The R-type, with a 600cc air-cooled Ransomes engine and lever-operated clutches to control the drive to the rear rollers and 36in cutting cylinder, weighed 9cwt. The three-section rear roller had a double differential to facilitate cornering.

The two-wheeled Overgreen greens mowers introduced in 1937, and made until 1958 under licence from The Worthington Mower Co., enabled a man to cut up to eighteen golf greens in a six- to eight-hour day. Self-propelled and pedestrian-controlled, the Overgreen had a 348cc single-cylinder four-stroke Ransomes engine, centrifugal clutch, large-diameter wheels with balloon tyres and a forward/neutral gear lever. An overall cutting width of 3ft 8in was achieved with one leading and two trailing 16in cut Certes fine turf cylinders, with individual grass boxes and mounted on flexible linkages to give accurate cutting on unlevel surfaces. The Overlawn 30in gang mower for cutting around trees and other areas of rough grass inaccessible to wider gang mowers was virtually the same as the Overgreen greens mower apart from a gang mowing cylinder in place of three fine turf cylinders.

The onset of World War II brought lawn mower production at the new White City works to a virtual stop. A few gang mowers were made for cutting airfields but most of the space was used to make 800 machine tools, 300 bomb trolleys and 1,500 trailers for 25-pounder guns. Other wartime production included thousands of parts for Bren-gun carriers, Crusader tanks, Rolls-Royce aero-engines and Mosquito aircraft.

Ransomes returned to lawn mower production on a reduced scale at the White City in 1945. A shortage of materials limited the production of four-stroke motor mowers during the immediate post-war years when Ransomes reintroduced their 16, 20 and 24in models with a 248 or 348cc engine and the 30in

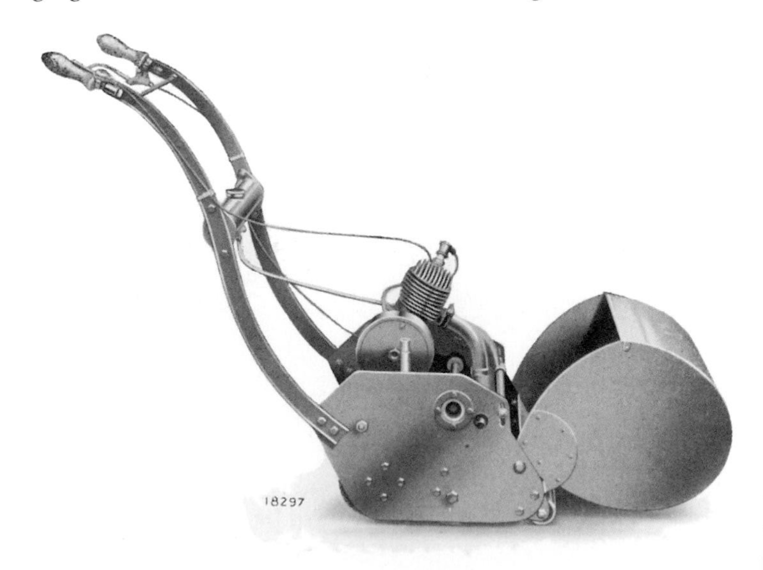

A car-type starting handle was used to start the Villiers 98cc two-stroke engine used on the 12 and 14in cut Minor motor mowers.

machine with a 600cc power unit. The 348cc Ransomes four-stroke engine was also used for the Verge Cutter and the Overgreen mower. The 40in dual-purpose mower and roller, first made in the early 1930s, was given a facelift with a new 10hp Ford power unit, additional guards and a cowling over the engine. The side-wheel Ripper, introduced in 1946 for cutting long grass, was one of the first new hand mowers introduced during the immediate post-war years, and the Mk II Ripper, which replaced it in 1963, was made for another eleven years.

The Astral, Mk III Ajax, Celer and Certes hand mowers; the Ariel, Moon,

The Manimal mower for man or animal draught was supplied with a draught rope and handle.

The Ransomes Verge Cutter.

The Antelope.

Tiger and Ripper side-wheel machines, and the 20 and 24in cut Manimal mowers were all listed in Ransomes' 1950 catalogue. The Manimal had replaced the earlier range of pony and bullock mowers and the Patent 24in Chain and Gear Automaton machines. As the name suggests, it could be used by two men, one pulling and one pushing, or pulled by a pony, donkey or bullock and probably steered by the head gardener. The Celer was a 14, 16 or 18in cut hand mower with an eight- or ten-knife cylinder for cutting golf and bowling greens or first-class private lawns. It had totally enclosed machine-cut steel gears, self-aligning ball bearings and two-piece aluminium rollers.

Motor mowers at the time included the 12 and 14in cut Ransomes Minor mowers (photo, page 104) and a 16in cut machine with a 147cc Villiers two-stroke engine. A 98cc two-stroke Villiers engine provided the power for the hand-propelled Auto-Certes fine turf cylinder mower and the side-wheel Gazelle, made between 1948 and 1959 for cutting long grass. The Clipper Motor scythe, introduced in the late 1940s, was a 'robust machine designed for work beyond even the capability of the Verge Cutter'. The 39in wide cutter-bar machine had a 348cc Ransomes four-stroke engine with dry sump lubrication, cone steering clutches in the pneumatic-tyred wheel hubs and a dog clutch in the cutter-bar drive.

The side-wheel Antelope and Auto-Certes motor mowers were launched in 1956. The Antelope, which replaced the Gazelle, had a 119cc B.S.A. four-stroke engine and a five-knife cylinder used for cutting rough grass. The Antelope was improved several times during its 37-year production which came to an end in 1993. Although only the cutting cylinder was driven on the Auto-Certes, sales literature

The Clipper motor scythe, designed to cut grass, weeds, bracken, brambles, etc., of any height, was especially constructed for easy operation.

explained that the mower was easy to push and that if the user changed walking speed, the number of cuts per inch could be varied to suit conditions. A self-propelled version of the Auto-Certes was introduced in 1958 and this mower was still in production in 2001.

The Mercury, Marquis and Sprite motor mowers were made in the late 1950s and early 1960s. Sales literature for the Sprite declared that this 14in cylinder mower, with a two-stroke J.A.P. engine, cut 500 square yards of grass for one penny. The 16in Mercury with a 75cc Villiers four-stroke engine was made between 1958 and 1965. The 18 and 20in cut Marquis became a firm favourite with professional users and private gardeners and has remained in production to the present day.

Introduced in the mid-1950s, the 18in Cyclone with a 2hp J.A.P. two-stroke engine, was the first Ransomes rotary mower. The two cutter blades were attached to a 'hummock disc' with a convex guard on the underside. Sales literature claimed that with this guard the cutting disc would ride over uneven ground without damage and long grass could not wrap around the disc shaft and possibly stall the engine.

The Cellec 18in cut mains-electric bowling green mower with a 240 volt ½hp motor driving the eight- or ten-knife cylinder was introduced in the late 1940s. It had an automatic centrifugal clutch and used half a unit of electricity to cut half an acre of grass in an hour.

Ransomes' 20 and 24in cut Electra lawn mowers, 'eminently suitable for all lawns and bowling greens', were still being made in 1950. There were two models. One had wooden front rolls, a standard bottom blade and a seven-knife cylinder for lawns. The other, with steel front rolls, a thin bottom blade and a ten-knife cylinder, was recommended for bowling greens.

The Cellec had reached the Mk 7 and Mk 7A stage in 1960. Improvements included 75 yards of cable and a chain-driven rear roller with a secondary clutch to disengage the drive to the cutting cylinder so that the mower could also be used to roll the green. The 1hp Mk 7A Cellec was, apart from its more powerful motor, identical to the ½hp Mk 7.

The range of Ransomes' hand and motor mowers for the private garden was in decline by the mid-1960s and within ten years the company had withdrawn from the domestic lawn mower market. The Conquest (photo page 111), Ajax and Certes roller mowers and the side-wheel ripper were the only hand mowers in the 1964 price list. The 12in cut Ajax with grass box cost £12 5s 6d. There was a choice of a B.S.A. engine or a battery-powered electric motor for the 14in cut Ransomes Fourteen cylinder mower made between 1964 and 1970. The 16in cut Mercury, the 18 and 20in cut Marquis motor mowers and the 14in Ripper side-wheel model completed the list of mowers. Gardeners preferring a rotary mower had the choice of the 18in side-discharge Typhoon with a two- or four-stroke engine, which replaced the Cyclone in 1960, or the rear-discharge Ransomes Rotary with a grass collector (photo page 110).

Ransomes dropped out of the domestic mower market in the mid-1970s and took no further interest in this sector until they acquired G.D. Mountfield of Maidenhead in 1985.

Ransomes' professional mowers in the mid-1960s included the 18in Auto-Certes, 24in cut Matador, 30 and 36in Mastiff, the Multi-Mower, the ride-on Motor Triple and five sizes of tractor gang mower. The two-speed Multi-Mower, introduced in 1960 for large areas of grass, was a dual-purpose machine with a 288cc J.A.P. four-stroke engine and interchangeable 30in cylinder and 27in rotary cutter heads.

Marquis motor mowers have been made for forty years and were still current in 2001.

Operation Matador

or the Adventures of a Ransomes Motor Mower

Four students at Hatfield Technical College approached the Lawn Mower Sales Department in 1959 with the idea of driving a Ransomes Matador lawn mower down the A1 non-stop from Edinburgh to London to prove the reliability of small petrol engines. The Lawn Mower Department Sales Manager, Bob Garnham, invited Mark Grimwade, who had just completed his special apprenticeship, to organise the project.

Ransomes were planning the introduction of the Pathfinder, a rough-grass version of the 24in Matador motor mower intended for council verge cutting with a three-knife cylinder, a pair of side wheels and rubber-coated rear roller. Two 'specials' were produced with trailer seats on pneumatic tyres and standard 288cc Villiers engines fitted with oversize sumps to hold extra oil. The rules did not allow for the engines to be stopped to check oil levels.

Mark and his team had no idea if the mowers would hold together for a non-stop 300-mile journey so it was agreed to carry out some 'in-house' trials on one of them. This involved four Ransomes apprentices driving the mower non-stop in the Ipswich area for ninety hours, which was the time calculated it would take to travel at 3mph down the A68 and A1. The new Nacton Road works were being built at the time and bunk beds were set up for the test team in the old army hut that was being used as the security lodge.

At night, testing was restricted to the roads around the works. Much to the amusement of the security staff, in the evenings numerous student nurses loyally supported the test team consisting of Keith Davidson, Dick Kellar, Tim Reeves and Jack Young. In the daytime the team could go anywhere within the range of their three-hour shift, and on one occasion one of them was seen cutting his girlfriend's lawn in a neighbouring village. The only breakdown occurred when one of the drivers stalled the engine while attempting to climb Bent Hill in Felixstowe.

With the trials successfully completed, the date of 'Operation Matador' was set for Easter 1959. Mark Grimwade set off for Edinburgh in a works Austin A40 Pick Up truck with two mowers on board (one as a standby) and Don Chipperfield took the Ransomes Film Unit movie cameras. The four Hatfield students had arrived at Edinburgh Castle in a Bedford Dormobile, which was to be their mobile home for the next four days and nights. The gift of a haggis was placed in the grass box for the Scots 'Keeper of the Royal Parks' in London and the long-distance lawn mower run set off to the accompaniment of a lone bagpiper.

Following as straight a line as possible the team went down the A68 over Carter Bar and across the border on a cold, dark Easter night to Scotch Corner, then down the A1 to London. This was long before the advent of dual carriageways and the convoy of a 3mph mower, Dormobile and A 40 Pick Up truck caused considerable congestion, particularly as the A1 still ran through the centre of most towns. As well as keeping a check on and re-fuelling the mower and its drivers, Mark had to roar off to the next town to alert the press, mayor and local Ransomes dealers of the imminent arrival of this wondrous sight. Appropriately the reception, ably assisted by Ransomes' outside staff, usually took place either in or just outside a coaching inn!

A three-hour shift in the cold weather on a bucking and roaring motor mower with no springs was about all the driver could stand. Then it was time for a fry-up and an attempt to sleep in the Dormobile as it crawled south. The mower ran like a dream and, like the test machine, appeared to burn no oil, which was just as well as there was no way of checking the level on an engine which was not permitted to be stopped. Four days and three nights later a high-powered reception committee of dealers, Royal parks staff, Ransomes' top management, television cameras and the press were waiting to meet the mower as it chugged its way through the London traffic and into Hyde Park dead on time. The haggis, by then somewhat soaked in petrol, was handed over and a ceremonial strip of Hyde Park was mown to Ransomes' perfection.

Don Chipperfield had filmed all the way down the A1 and was determined to complete his epic 'Operation Matador' with a splash. Tim Reeves was called into service again and dressed as a yokel in a smock with straw in his hair. He was given an old bicycle and told to 'act' in amazement on seeing the approaching mower and ride it straight into a duck pond. This he did to perfection. The film was shown for many years and the bike is probably still in that pond!

Heading for London on 'Operation Matador'

Improved Mk II and Mk III Multi-Mowers were announced in 1961 and 1969 respectively. The Multimower 2000 (now without a hyphen in the name) was made between 1972 and the mid-1980s when it was replaced by the 30in Multimower Reelcutter, a machine still current in 2001.

The triple, quintuple and septuple trailed gang mowers introduced in the 1920s were still in production in the mid-1960s, as was the nonuple gang mower launched in 1939 with a complicated hitch system linking the nine mower units together. There were two types of cutting cylinder driven by steel or rubber-tyred wheels. The six-knife Sportcutter was recommended for grass up to 5in high and the five-knife Magna was advised for cutting grass up to 8in high.

Moving trailed gang mowers from site to site, either on a trailer or hitched in crocodile formation behind a tractor or road vehicle, was time consuming. Professional users were looking for high-output gang mowers that could be transported at speed. Introduced in 1964, the Ransomes Power Quint, with five cutting cylinders, was the world's first three-point linkage mounted gang mower driven by power take-off. High-speed road travel was now possible and the two outer units could be lifted manually to a vertical position to reduce the overall width for transport purposes.

A 1965 sales leaflet priced the 12 volt battery electric Ransomes Fourteen at £52 10s. The four-stroke petrol-engined model was cheaper by £10.

Ransomes introduced the 18 inch Rotary in the early 1960s.

The Motor Triple ride-on three-unit gang mower with a Villiers four-stroke engine and the Junior trailed gang mower were also launched in 1964. The 86in cut Motor Triple, with one underslung and two front-mounted cutting cylinders, was the first machine of its type made in Britain and was made for the next twenty years or so. The original Motor Triple had five-knife cutting cylinders but within a year Ransomes had added a bank version of the Motor Triple with three-knife cylinders for cutting rough grass, and an optional M.A.G. diesel engine was now available for both models. The three- and five-unit

The stylish 12in cut Conquest cylinder mower and the Ajax were the only Ransomes hand mowers made in 1965 for the smaller garden. The Conquest cost £7 11s and the Ajax was £12 5s.

Junior gang mowers, designed for use with 7 to 10 hp four-wheel tractors, had cutting widths of 56in and 92in respectively. They ran on solid rubber-tyred wheels and were hitched in crocodile fashion for transport.

The 15ft cut Hydraulic 5/7 hydraulically driven tractor-mounted gang mower introduced in 1967 was another world first for Ransomes. The Hydraulic 5/7 was easily wrapped around a Fordson Dexta or Massey Ferguson 135 tractor and the five or seven cutting units were independently driven by hydraulic motors. The driver was able to raise or lower the cutting units and engage or disengage the drive without leaving the seat. The 5/7 could cut up to ten acres an hour at 7mph and large numbers of these hydraulically driven gang mowers were sold in America and Canada.

The seven-unit machine had a 15ft cutting width, but as each cylinder was independently driven it was possible to use only the number of units sufficient for the width of grass to be cut. It was hydraulically folded

The Multi-Mower power unit could be used with a 30in cut cylinder mower or 27in wide rotary cutter. This machine has the necessary guards to comply with farm safety regulations which came into force on 1st July 1964

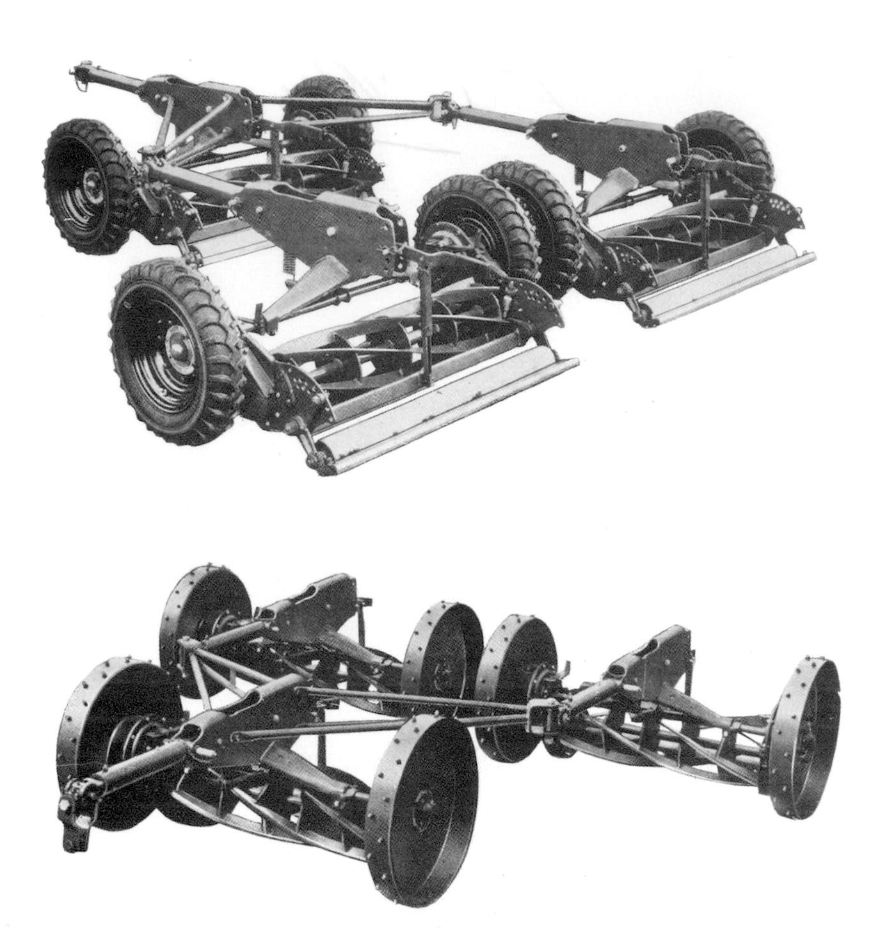

Sportcutter and Magna gang mowers cut up to three acres an hour at 5mph.

The Junior gang mower was designed for use with 7 to 10 hp mini-tractors that were popular in the mid-1960s.

to 8ft 2in for transport. However, because two of the cutting cylinders on the 5/7 were mounted under the tractor, it took a fair amount of time to remove the mower from it. Introduced a year or so later, the mounted Hydraulic 5 for cutting golf course fairways and the 7ft cut Hydraulic 214 for roadside verges, sports fields and housing estates could be attached to a tractor three-point linkage in about five minutes. The five independent, hydraulically driven, floating six-knife cutting cylinders on the Hydraulic 5 had a work rate of up to five acres an hour at 5mph. The Mounted 214 with three floating-head six-knife Sportcutter cylinders or the same number of fixed four- or six-knife Magna cutting cylinders had an output of three acres an hour at 5mph.

Ransomes' trailed gang carrier was another late-1960s introduction. Towed from the tractor drawbar and coupled to its hydraulic system, it had seven cutting cylinders driven by individual hydraulic motors. Hydraulic rams raised and lowered the gangs collectively or in any grouping from one to seven and the whole unit folded down to 8ft 2in for transport.

The three-unit Hahn Tournament Triplex ride-on greens mower (photo, page 116) was added to the Ransomes range in 1971. The three nine-knife cutting cylinders with large-capacity polyethylene grass boxes had an overall cutting width of 67in. A 12 hp Kohler engine provided the power for the hydrostatic transmission and the cutting cylinders were driven by a combination of vee-belts and flexible cables. The quick-attach cutting cylinders could be replaced with Verti-cut thatch removal reels or vibrating spikers for aerating the turf. Hahn greens mowers were painted yellow but the similar Ransomes Tournament Triplex 171, made under licence at Ipswich from 1972, had the more familiar Ransomes green paintwork.

At £113 and £133 respectively the 18 and 20in cut Marquis were the cheapest of Ransomes' motor mowers in the early 1970s. The 24in cut Ransomes Twenty-Four motor mower cost a little over £200 but a trailing seat and electric starting were extra. The 24in Matador (photo, page 117) with a cast-iron land roll was £278 or £293 with a rubber-covered roll. The most expensive model in Ransomes' price list, the Mastiff, with a 30 or 36in cylinder and optional self-emptying grass box, had a price tag of about £500. The Auto-Certes with an 18 or 20in cut ten-knife cylinder cost from £160 to £189 depending on specification. It had a four-stroke engine and separate clutches for the land roll and cutting cylinder. Detachable power-driven transport wheels, 'saving tedious pushing between greens', were an optional extra.

A growing proportion of Ransomes' total output was provided by the grass machinery division in the early 1970s and an increased demand for turf maintenance

Ransomes' Motor Triple mowers had a Villiers petrol or M.A.G diesel engine.

The Hydraulic 5/7 mounted gang mower.

The Daily Mail Air Race

Flushed with the success of 'Operation Matador', Mark Grimwade and his team decided that for an encore they would take part in the Daily Mail's Marble Arch to the Arc de Triomphe Air Race, held in July 1959 to celebrate the fiftieth anniversary of Blériot's first cross-Channel flight. The rules were simple: whilst breaking no laws, the competitors had to get to an airfield by any possible means and fly across the Channel. The fastest time won.

The team geared up the land roll drive on both Matadors to increase cruising speed from 3 to 15mph. Matador No.1 hurtled off to Croydon airport where Ransomes had hired the oldest Tiger Moth still flying. The driver was airlifted without the mower to a small airport just outside Paris, where Ransomes' French dealer M. Perrier was waiting with Matador No.2. However when they increased the Matador's top speed to 15mph, the team had not allowed for the effect of

Leaving Hyde Park en route to Croydon airport

France's cobble-stoned streets, and before long the fuel tank fell off! M. Perrier took the team for a five-course lunch with more wine than Perrier water while his mechanic fixed the Matador with elastic. The 'Matadors' completed the course in 10 hours and 48 minutes. The winner, in a military style operation, took just 40 minutes 40 seconds using a motorcycle, a helicopter from a Thames pontoon to Biggin Hill and a Hawker Hunter to get to Paris. Another helicopter took the rider to the Champs Elysées Avenue to complete the last few yards to the Arc de Triomphe on a second motor cycle

Daily Mail

£10,000

BLERIOT ANNIVERSARY RACE

LONDON-PARIS

JULY 13-23 1959

This is to certify that

Mr. M. Grimwade

competed in this event sponsored by the Daily Mail and organised by the Royal Aero Club of the United Kingdom to mark the 50th anniversary of M. Blériot's historic first flight across the English Channel.

Mark Grimwade's Air Race certificate proves the team really did arrive at the Arc de Triomphe

equipment for public parks and sports grounds saw sales of grass machinery exceed those of farm machinery for the first time in 1973. A new high-output, hydraulic-drive Motor 5/3 ride-on mower announced in 1974 was another first for Ransomes. As well as driving the cutting cylinders, hydraulics were used to lift the units from work, fold them for transport and provide an infinitely variable speed control mechanism for the transmission system. Following the success of the Hydraulic 5/7 tractor gang mower (photo, page 113) the Hydraulic 3, a similar three-gang tractor-mounted model for small- and medium-sized areas of grass, was launched at Motspur Park in 1975. The 7ft cut, mounted 214 Hydraulic Gang Mower added in 1979 had three hydraulically driven Magna four- or six-knife fixed-head cutting cylinders or six-knife floating-head Sportcutter cylinders.

Ransomes introduced the American Hahn Tournament Triplex mower to British sports grounds in 1971.

Ransomes imported Bob-Cat professional rotary mowers made by Wisconsin Marine Inc. of Milwaukee for the first time in 1977. In the following year they acquired a thirty per cent stake in Wisconsin Marine to gain a stronger foothold in the American and Canadian grass machinery markets. There was a Bob-Cat machine for every possible application ranging from a 12in strimmer to a 74in cut ride-on rotary with a 20hp engine. Wisconsin Marine's Model 61 Rider Rotary was added to the range of Ransomes' grass machinery in 1978. The specification included a twin-cylinder 16hp petrol engine, hydrostatic transmission, front-wheel drive and rear-wheel steering. The fully floating 61in cutter unit had three 21in long cutting blades suitable for cutting grass up to 9in high.

Ransomes were the major shareholder in Wisconsin Marine by 1980 when their catalogue included 48, 61 and 74in cut Rider Rotaries with a 20hp Onan twin-cylinder petrol engine. The new all-hydraulic Motor 213 and Motor 180 self-propelled triple-cylinder mowers appeared in 1981. The Motor 213 with a twin-cylinder Kohler petrol engine and three cutting cylinders with a 7ft working width could cut up to four acres an hour at speeds of up to 6mph.

The Motor 180 was a joint project between Ransomes, Ipswich and Ransomes Inc. in America (formerly Wisconsin Marine), who provided the power unit for a fixed or floating cutter head made at Ipswich. Features of the two-acres-an-hour Motor 180 included an 11hp Briggs & Stratton single-cylinder engine, hydrostatic transmission and an over-centre clutch on the drive belt to the cutting unit. A later version, designated the Motor 180D, had a twin-cylinder water-cooled Kubota diesel engine and the choice of fixed or floating cutting cylinders.

The Motor 350D all-hydraulic ride-on gang mower

with five cutting cylinders and a 38hp Kubota diesel engine was added in 1983. It had hydrostatic steering on the rear wheels; a single pedal provided forward and reverse directions of travel through a hydrostatic transmission, and it was supplied with fixed or floating cutting cylinders. Improvements in 1989 included a quieter diesel engine and variable speed control for the cutting cylinders.

Ransomes were still making a full range of professional rotary and cylinder mowers in the late 1970s and 1980s. The Marquis, Twenty-Four, Matador and Mastiff motor mowers were suitable for the upkeep of formal gardens and tennis courts, while the Auto-Certes was for fine turf. There were various pedestrian-controlled rotary mowers, including the Antelope, Verge Cutter and Multimower, for cutting verges and small areas of rough grass. Professional machines included numerous ride-on rotary mowers and triple-gang mowers for long grass, the Motor 5/3 gang mower and a wide selection of hydraulic motor or ground-wheel-drive mounted and trailed gangs for sports ground maintenance.

163 The rear-wheel steered Motor 180 had an 11hp Briggs & Stratton engine and a hydrostatic transmission with a top speed of 7mph.

A Matador mower was driven non-stop from Edinburgh to London in 1959 (see page 108).

Ransomes returned to the domestic lawn mower market in 1985 when they bought G.D. Mountfield Ltd of Maidenhead. Following the purchase of Westwood Tractors at Plymouth in 1989, the two companies were merged to form Ransomes Consumer Ltd at Plymouth.

G.D. Mountfield Ltd was established in 1962 to manufacture lawn mowers and garden cultivators. Advertisements for the Mountfield M1 cultivator introduced in 1967 described it as the ultimate in garden mechanisation. Powered by a 3½hp Briggs & Stratton engine, the M1 was a dual-purpose

An improved Verge Cutter, introduced in 1981, had a vee-belt drive from the M.A.G single-cylinder petrol engine to the two-speed gearbox as well as a 30in four-knife reel cutter.

machine that could be used with a rotary cultivator or lawn mower attachment. Mountfield also made the first pedestrian-controlled petrol-engined mower with electric starting.

Westwood started out in 1969 as a small manufacturer of pedestrian rotary mowers and they introduced the first British-built ride-on garden tractor, known as the Lawnbug, in 1977. Later products included a range of ride-on lawn and garden tractors with mid-mounted rotary mower decks.

Many machines with names familiar since the early 1980s were included in the 1992 Ransomes catalogue. The Antelope, Marquis, Matador, Mastiff and Super Certes motor mowers were listed together with various models of ride-on, mounted and trailed tractor gang mower and the Motor Triple.

Scores of new golf courses were being built in the early 1990s to meet the needs of the increasing number of British people taking up the game. In residential areas where golf courses were in great demand it was necessary to cut the greens and fairways very early in the morning, but the noise of a petrol- or diesel-engined ride-on mower at this hour was not popular with local people. This problem was solved in America in 1994 when Cushman & Ryan launched the world's first electric greens mower.

The Ransomes 62in cut E-Plex all-electric greens mower resulted from a joint project using Cushman's electric vehicle expertise and Ransomes' vast experience of greens mowers.

The Mountfield Mirage hand-propelled 31/2 hp rotary mower.

The 1995 range of walk-behind Bob-Cat rotaries with hydrostatic drive model with a 30 or 48in cutter deck.

A 48 volt 2hp motor provided infinitely variable forward and reverse drive, and a ½hp motor was coupled directly to the three nine- or eleven-knife floating cutter heads. With a fully charged set of eight 6 volt batteries, the E-Plex had maximum cutting and transport speeds of 4 and 7½ mph respectively and, depending on course layout, it could cut between nine and eighteen greens in three hours.

By the late 1990s a range of more than thirty Ransomes, Cushman and Ryan grass-cutting machines including ride-on and tractor gang mowers, rotaries, triple-reel mowers and walk-behind motor mowers were being made in the UK, America and France and sold in all parts of the world. The motor mowers included the latest models of the Mastiff, Marquis, Matador, Super-Certes (first made in 1987) and pedestrian-controlled Rotary 45 with an optional grass catcher. There were Bob-Cat and Front Line rider rotaries, electric E-Plex and petrol- or diesel-engined T-Plex triple-reel mowers, as well as ride-on gang mowers for cutting golf course fairways. Tractor gang mowers included the Hydraulic 5 and 5/7, five- and six-unit trailed models and the five- or seven-unit all-hydraulic trailer-mounted gangs for large open areas of grass.

The Ransomes E-Plex electric greens mower, launched in 1994, uses no fuel or oil which might leak on to fine turf; its quiet operation makes it ideal for use on golf courses close to residential areas.

Developed from the Bob-Cat by Ransomes Inc. at Wisconsin, the Front Line 700 series ride-on rotary mower with a Kubota diesel or Kohler petrol engine.

Turf Equipment

Ransomes were making leaf sweepers, rollers and lawn edge trimmers in the early 1930s. The edge trimmer, with six rotating knives cutting against a fixed blade, cut grass edges as fast as a man could walk. The working depth of the knives could be adjusted to suit the type of lawn edge and the knives were easily replaced when worn. Ransomes' 24 and 36in leaf sweepers were described at the time as an 'excellent labour saving device for gathering up leaves and light rubbish from well kept lawns in a more economical and expeditious way than is possible when sweeping by hand.' A revolving brush, driven

Ransomes all-hydraulic trailer-mounted gang mower.

by side wheels, swept the leaves into the canvas collector and the action of the brush was said to have a beneficial effect on the grass.

Ransomes' Rolezi and Automaton garden rollers were double-cylinder models with roller bearings. The Automaton was of the very latest design, and adjustment was provided to take up wear between the two halves of the roll. The Orwell double-cylinder and a cheaper single-cylinder garden roller could be filled with water for added weight. The largest 30in diameter and 30in wide Orwell roller weighed just over 10cwt empty and almost 15cwt when filled with water. Specialist rollers were made for bowling greens and hard tennis courts. The single-cylinder bowling green roller with roller bearings and an extra long handle was 'designed by a leading expert and a lad could use it with ease'. The twin cylinders on the hard-court roller were specially turned in a lathe to make them suitable for the work. A water-ballast version was made for asphalting and municipal work.

This 24in wide leaf sweeper, made at the Orwell Works in the early 1930s, had a £9 price tag.

Supreme Mowing Ltd, who made mower cylinders and cylinder grinders, became part of the Ransomes group in 1987, and in the following year Ransomes acquired Steiner Turf Equipment Inc. of Dalton, Ohio, who manufactured two- and four-wheel-drive tractors and turf care attachments. Two more North American companies, Cushman & Ryan and Brouwer Turf Equipment Ltd, joined the Ransomes group in 1989.

Steiner had introduced a small tractor in 1975 and, following Ransomes' acquisition of the business, Steiner developed the Turftrak system based on a rigid two-wheel-drive and articulated four-wheel-drive turf tractors. A wide range of attachments for the Turftrak

The blades on Ransomes' 1930s lawn edge trimmer could be raised clear of the ground when moving the edger from place to place.

A new Ransomes leaf blower introduced in 1981 and made by Ransomes Inc. in America had an 8hp Briggs & Stratton engine to drive a high-velocity fan. The leaves were blown into rows or heaps for collection.

included three sizes of front rotary mower, triplex reel mower, turf aerator, disc edger, bunker rake, lawn sweeper and power blower. The two-wheel-drive Turftrak 2 with a 16hp water-cooled diesel engine, hydrostatic transmission, diff-lock, rear-wheel steering and independent brakes was discontinued in 1991. The four-wheel-drive Turftrak 4, with an optional 21hp diesel engine, remained in production until 1995.

Brouwer, with premises near Ontario, were founded in 1972 to manufacture the world's first automated turf harvester followed by a range of turf maintenance equipment. Ransomes relocated Brouwer in Ohio to share production facilities with Steiner.

Cushman started making two-stroke farm engines at Lincoln, Nebraska in 1900, and over the years the two Cushman brothers widened their interests to include vehicles for grounds maintenance and light industrial use. The load-carrying Turf-Truckster with its range of turf care equipment for golf courses was developed in the early 1960s. The Ryan business was established in Minnesota in 1946 to make turf cutters, and pedestrian-controlled and ride-on turf renovation equipment had been added by 1969 when they became part of the Cushman business. Production of Ryan turf equipment was re-located to Lincoln, Nebraska in 1980.

Ransomes' acquisition of Cushman & Ryan added three- and four-wheel diesel-engined Cushman Turf-Trucksters and petrol-engined Junior Turf-Trucksters together with Ryan ride-on and pedestrian-controlled aerators and turf cutters to their range of turf care equipment.

Ransomes' Green Machines, introduced from America in 1981, were a group of three nylon cord grass trimmers with 17.2 to 37.4cc two-stroke engines. Brush cutter, saw and pruning heads could also be used with a Green Machine.

Steiner made the Turftrak in America. Attachments included a front-end loader, disc edger, turf aerator, lawn sweeper and front-mounted triplex reel mower.

The Turf-Truckster was a dual-purpose machine which, as well as being a handy transport runabout with a top speed of 22mph, could also be used with various turf care attachments including an aerator, top dresser, sprayer, fertiliser spreader and core harvester. The E-Plex all-electric greens mower was the most significant new product of the 1990s and production of a full range of Ransomes, Cushman and Ryan turf care machinery continued when Ransomes became part of the Textron Corporation in 1998.

A Cushman Turf-Truckster with a core harvesting attachment collecting soil cores left by a Ryan aerator.

Chapter 7
Sprayers

Crop sprayers were a relatively late addition to Ransomes' range of farm machinery. Horse-drawn spraying machines, usually with a stationary engine to drive the pump, were made by other companies in the 1930s. Ransomes' involvement in crop protection began in the early 1940s with the introduction of the Agro atomiser sprayer, allegedly developed when it was feared that enemy aeroplanes might drop Colorado beetles on British potato fields in an attempt to destroy the crop.

The Agro sprayer, designed in conjunction with the I.C.I. Engineering Research Department at Billingham and manufactured under licence from Plant Protection Ltd, was one of the first low-volume sprayers made in Britain. Working at 5psi, it applied ten gallons of spray chemical per acre either to row crops or as an overall treatment. The chemical, mixed at ten times the usual strength, was introduced into a stream of air created by a large fan, which atomised the liquid, and applied it through two pairs of nozzles on a set of pendant legs.

There were three versions of the Ransomes Agro sprayer: trailed, close coupled to the tractor drawbar, or mounted on the three-point linkage. The tractor-drawn sprayer had a 120 gallon tank, and thirteen pendant legs spaced for 28in rows gave an overall span of 31ft. The centrifugal liquor pump and fan were vee-belt driven by a 20hp water-cooled four-cylinder engine. The spray booms were folded for transport by a system of ropes connected to a vertical hydraulic ram supplied with oil by a vee-belt driven hydraulic pump. The pendant legs were attached with a simple bayonet joint and each leg had two upper and two lower atomiser nozzles attached to independently adjustable ball and socket joints. The lower nozzles were set to spray upwards into the foliage when working in row crops and all four nozzles were angled down for overall spraying. The nine pendant legs on the close-coupled Agro sprayer gave a spraying width of 20 ft, the tank held 75 gallons and the wheel track was adjustable in 1in steps from 4ft 4in to 5ft 4in. The sprayer was set close to the tractor with the power take-off shaft driving the liquor pump and fan. The driver was able to pull either side boom close up the tractor rear wheel to avoid hitting the hedge when making tight headland turns. The side booms were folded forward for transport and a longer drawbar was provided to set the sprayer further away from the tractor to allow unrestricted manoeuvring in traffic.

The nine-row mounted Agro sprayer was similar to the close-coupled model but it had two 40 gallon liquor tanks saddle-mounted on each side of the tractor, and the side booms were locked alongside the tractor for transport. All three models were equipped with an acid-resisting liquor pump, self-filler

The spray booms on the nine-row tractor-mounted Agro sprayer with an overall spraying width of 20ft were folded down to 7ft for transport.

Small quantities of insecticide dust were applied with the Ransomes Flea Beetle duster. Introduced in the late 1940s, the hopper held 10lb of chemical, and a wheel-driven rotating brush distributed the powder along the rows.

attachment, a water meter indicating the quantity of liquid applied and a control valve to regulate the flow of air from the fan.

The Mk 1 Cropguard trailed sprayer, suitable for use with most popular makes of tractor, replaced the Agro sprayer in 1949. Mounted and trailed versions of the Junior, Standard and Senior sprayers with hot-dip galvanised cylindrical steel tanks replaced the original Cropguard sprayer in 1955. Prices started at £55 for the mounted Junior model and rose to £152 for the trailed Senior with a 20ft spray boom. The 30 gallon low-volume Junior with a roller vane pump and a pressed-steel, three-section spray bar applied 5 to 20 gallons per acre with a 17 or 20ft spray bar. The medium- to low-volume Standard with a 50 gallon tank, a 21 or 24ft pressed-steel, three-section spray bar and a gear pump had an application rate of 5 to 35 gallons per acre. There were two versions of the 100 gallon Senior: the Mk 4 was a low-volume sprayer with an application rate of 5 to 35 gallons per acre, and the low-high volume Universal applied between 5 and 100 gallons per acre at 4mph. Both had a 20ft or optional five-section 32ft angle-iron spray bar and the Universal sprayer had a high-capacity gear pump.

All Cropguard sprayers had flat fan jets; the pump was mounted directly on the power take-off shaft, and a break-back mechanism was built into the spray boom. Accessories included canvas wind sheets to reduce drift, rake-head and single nozzle hand lances

The Ransomes Cropguard Junior, a low-priced sprayer for small farms, was advertised as particularly suitable for weed control in cereal crops and grassland.

No-drift hollow-cone nozzles were used on the FR Cropguard precision band sprayer.

and a self-fill attachment. Other optional equipment for the Cropguard Senior included pendant legs for spraying potatoes and a mechanical paddle agitator required when applying chemicals that remained in suspension. The paddle was driven by belt from a land wheel on the trailed model and from a rear tractor wheel on mounted machines when used with David Brown, Ferguson, Fordson and Nuffield tractors.

The 50 gallon Cropguard Mk 4 trailed sprayer with a 20ft spray boom was made for the MG Motor Cultivator. It was similar to Ransomes' farm sprayers with a gear pump driven by the power take-off shaft and, depending on the size of jets fitted, the application rate varied from 11 to 51 gallons per acre at the MG's top speed of 2½ mph. Hand lances were available for spraying fruit bushes and trees. Trailed sprayers were made by A & G Cooper of Wisbech, W. Weeks & Son of Maidstone and Drake & Fletcher. The Cooper Demon fruit sprayer had a 100 gallon tank and two hand lances, and the pump was mounted on the MG and chain-driven from the power shaft. The pump and 50 gallon tank were mounted on the chassis of the trailed Weeks Model M fruit sprayer which could be used to spray ground crops or soft fruit. Hand lances were provided for spraying fruit trees. The Drake & Fletcher L.O. Estate sprayer with a 60 gallon trailed tank and power take-off driven pump mounted on the

Up to fifteen acres could be sprayed in an hour with the Cropsaver sprayer.

tractor could be used with a spray bar or hand lance for spraying ground crops, fruit trees and hops.

Mk 2 FR Cropguard Junior and Standard and Mk 5 Cropguard Senior sprayers were announced in 1961. Compared with previous models, the Mk 2 Junior had a 50 gallon tank, and the tank capacity of the Mk 2 Standard was increased to 60 gallons. New shut-off taps for the outer sections of the spray boom, priced at £2 a pair, were added to the list of optional extras. Strengthened spray boom extensions and modified pump mountings for the latest tractors were the main changes on the Mk 5 Senior sprayer which was superseded by the FR Cropguard 100 in 1964. It was similar to the Senior sprayer with a 20 or 32ft spray boom and designed for use with the new Basildon range of Ford tractors. Mk 3 versions of the Junior and Standard Cropguard sprayers appeared in 1965.

The FR Cropsaver 100 sprayer with a 40ft wide front-mounted spray bar also appeared in 1964. It had the same pump, filters and controls as the Cropguard 100, and the spray bar, manufactured under licence from Cleanacres Ltd at Cheltenham, could be attached to any make of tractor front-end loader. The 100 gallon tank was carried on the three-point linkage and the high/low volume pump mounted on the power take-off applied the chemical at rates of 7 to 40 gallons per acre. Sales literature explained that the risk of the operator being affected by chemical drift was no greater than when using a rear-mounted sprayer and the driver was able to immediately spot a blocked nozzle or adjust spray-bar height to suit varying ground contours. The Cropsaver 150 with a 150 gallon tank was added in 1966.

Band spraying was fashionable in the early 1960s. The Ransomes FR band sprayer was suitable for use with most popular makes of precision seeder and it had the capacity to band spray between four and ten rows at once. The chemical was carried in a 30 gallon tank mounted on the front of the tractor or in a 30, 50 or 60 gallon tank mounted on the seeder unit toolbar. The chemical was applied at very low pressure and it was important to apply an equal dose of herbicide to each row to achieve effective weed control. This was done by using factory-matched sets of nozzles and a diaphragm pressure-reducing valve to maintain constant spray pressure.

The Mk 3 Junior went out of production in 1967, and the Mk 3 Standard was made for another two years. From 1966 customers concerned about the hazards of spray drift could specify low-pressure cone nozzles or Vibrajets for Cropguard 100 and 150 sprayers. Developed by Plant Protection Ltd, the Vibrajet nozzle consisted of an electric oscillating mechanism, powered by the tractor battery, which applied large droplets of chemical at very low pressure. Various colour-coded Vibrajet nozzle sleeves provided different outputs and spray pattern widths. One type of Vibrajet suitable for the Cropguard 100 and 150 had a 6ft wide spray pattern while others with a narrow spray pattern were used on the Cleanrow chemical hoe.

Ransomes introduced the Cropsaver 150 in 1966, followed by the Cropguard 350 and Cleanrow sprayers in 1967. The Cropsaver 150 with its two-section 48 or 60ft aluminium boom mounted on hydraulic dampers and a 150 gallon galvanised steel tank had an output of up to 20 acres in an hour. Sales literature explained that the new boom design would eliminate bounce, and that 'feelers' at the outer ends would keep the boom level across its full width. The Cropguard 350, which was the first Ransomes trailed sprayer, had a 32ft spray boom, adjustable wheel settings, a 30 gallons per minute pump and a power filler which filled the 350

The Cleanrow inter-row sprayer could be steered from a seat on the machine.

Cropguard sprayers with corrosion resisting polythene tanks were introduced at the 1969 Smithfield Show.

gallon galvanised steel tank in twelve minutes. The Cropguard 350 tank and pump was also used with the front-boom Cropsaver sprayer.

The FR Cropguard Cleanrow inter-row sprayer or chemical hoe with Vibrajet no-drift nozzles, had independent floating steel guards to keep the chemical off the rows of sugar beet, turnips, strawberries, onions and other crops. The basic model was 9ft wide and the tool frame could be widened in 18in steps to 12ft. Depending on row spacing the Cleanrow could treat between five and twelve rows in a single pass. The chemical, carried in a 30 or 60 gallon front-mounted tank, was applied at a rate of 25 gallons per acre and upwards.

There were seamless polythene tanks, continuous in-tank liquid agitation and corrosion-resistant roller vane pumps directly mounted on the power take-off shaft for the new range of FR Cropguard sprayers announced at the 1969 Smithfield Show. The Cropguard 85 and 105 had 25ft 6in spray bars while the 105 Super and Cropguard 155 were equipped with a five-section, 31ft 6in wide spray bar. The model number indicated tank capacity, and the original roller vane pumps, direct-mounted on the power take-off shaft, were superseded by diaphragm pumps in 1972. Advertised as 'One of the Ransomes Land Army', the FR Cropsaver with a front-mounted 48 or 60ft aluminium spray boom was updated at the same time with a 105 or 155 gallon Cropguard Super polythene tank and diaphragm pump.

The last Ransomes band sprayers were made in the early 1970s when the more effective post-emergence sugar beet herbicides came into use, but the Cleanrow inter-row chemical hoe remained in production until

The Dorman Compact 1200 became the Ransomes Compact in 1981. It had a three-piston diaphragm pump mounted on the sprayer chassis, a polyethylene tank and a 10 or 12ft wide mechanically suspended spray boom.

Ransomes' Up Front tank unit held 550 litres of chemical. A hydraulically driven centrifugal pump transferred the liquid to the main tank.

the mid-1970s. Ransomes were still marketing the ageing FR Cropguard range when they acquired the Dorman Sprayer Co. in 1978. Manufacture of the Dorman mounted range continued at Ely but it was the Dorman Sprayer Co. at Nacton Works in Ipswich that published the sales literature. Production of the Ransomes Cropguard 85, 105 and 155 models was transferred to the Johnson's Engineering factory at March in 1978 when they became the Cropguard 400, 500 and 700; likewise the trailed Cropguard 350 became the 1600. The specification was improved with new diaphragm pumps and a mechanical suspension system on the spray bar. The new model numbers indicated tank capacity in litres.

Sprayer production was transferred to Ipswich in 1980. The Dorman name disappeared from Ransomes' sales literature in 1981 when the sprayer range included the trailed Compact 1200 and Super TD4. The Super TD4 had a fully suspended, self-levelling 12m spray boom, a four-cylinder diaphragm pump mounted on the chassis, a choice of a 1,600 or 2,000 litre glass-reinforced plastic tank and remote electronic controls in the cab. Ransomes' production records for 1981 included the Eagle, Hawk, Merlin and Falcon sprayers and an Up Front tank unit. The Eagle, with a 700 or 800 litre tank and 12m spray bar, was the top model in the new range of Ransomes' mounted sprayers. The Hawk was next in line with a 550 litre tank and 10m spray boom or 700 litre tank with a 12m boom. Both models had a hand winch to alter the height of the spray boom, cab-mounted electronic controls and a diaphragm pump. The control system for the Eagle also enabled the operator to vary forward speed in a given gear while automatically maintaining the pre-set application rate. The 200, 300 and 400 litre Merlin and Falcon with a 400 or 550 litre polythene tank and a three-section 6m or five-section 9m spray boom were aimed at the smaller farmer. The contents of Ransomes' front-mounted 550 litre polyethylene tank were transferred to the main sprayer tank with a centrifugal pump driven by a hydraulic motor. An in-cab electronic unit controlled the transfer pump, which could empty the front tank in three minutes. A similar 400 litre front tank outfit was tailor-made for use with

Ransomes' prototype CTV 2500 Crop Treatment Vehicle had a top road speed of 22 mph. The cab was tilted to provide access to the engine and the hydraulic pump.

The atomisers (below) on the Micronair sprayer were individually belt driven by small hydraulic motors supplied with oil from the tractor's hydraulic system.

the Hawk 550 sprayer on a Ford 1900 tractor. Use of a front tank was recommended for spraying winter-sown cereal crops using a tractor equipped with low ground pressure tyres. About a thousand Merlin, Falcon and Eagle sprayers and another thousand Hawks had been sold when the range was discontinued in 1984.

The Ransomes Micronair 700 air-assisted sprayer, launched in 1982, had seven rotary atomisers individually driven by hydraulic motors, on a fully suspended self-levelling boom. A fan in the rotary atomiser head created a draught of air that carried the chemical down into the crop canopy at a rate of 3 to 28 gallons per acre. An additional hydraulic power pack with a reservoir, pump and control valve was needed for tractors with insufficient oil flow from the hydraulic system, and an in-cab electronic control unit warned of malfunction of the rotating heads. Only twenty-three Micronair sprayers had been sold when the last one was made in 1984.

Ransomes spent two years in developing the CTV 2500 self-propelled sprayer which made its public debut in September 1981 with an anticipated price tag of £25,000. The pivot-steered Crop Treatment Vehicle with an 80hp Ford industrial engine and hydrostatic drive to all four wheels was the largest British-built low ground pressure sprayer in existence at the time. The three sections on each side of the 24m gimbal-mounted high-clearance spray boom were made from a glass reinforced plastic material. It was hydraulically folded for transport and other rams were used for height and tilt adjustment. Three CTV 2500 sprayers were eventually sold, and the design was included in the package when E. Allman & Co. at Chichester bought the Ransomes sprayer range in 1984.

Chapter 8

Root Crop Machinery

Potato Ridgers

Ransomes were making single-furrow ridging ploughs and ridging body attachments for single-furrow horse ploughs in the early 1800s. Wooden-frame ridging ploughs with an iron ridging body for teams of two or three horses or bullocks were made for the export market. A mid-1860s Ransomes catalogue listed five sizes of one- and two-horse improved iron ridging or moulding ploughs with two wheels. Ploughs with a cast frame cost from £2 12s 6d to £4 2s 6d, a wrought-iron frame added either 15s or £1 but between 5s and 7s 6d could be saved if the plough was supplied with a single wheel. The plough body with 'screw expanding breasts' was recommended for 'ridging or moulding up potatoes, beetroot and other plants sown on the ridge, and for opening water furrows'. When setting out land the marker saved 'the trouble of measuring or dividing the land before commencing work, as the plough, whilst making one furrow is marking a course for the next'. The catalogue explained that the ploughs could also be fitted with hoe blades and cutters for cleaning land between rows and an 'ordinary plough body for doing regular ploughing, and with shares and prongs for potato raising it will raise three to four acres in a day'. Farmers who felt unable to justify the cost of a ridging plough could buy a ridging body and a potato-raising share for the Newcastle series (Page 21) and other Ransomes ploughs.

Various sizes of wheeled ridging plough with hinged mouldboards adjustable for different widths of ridges were made in the 1920s and 1930s. The frame was of similar design to the YL mouldboard plough and could also be used with potato-raising and sugar beet lifting bodies. The Grampian swing ridging plough primarily designed for Scotland was supplied with large or small mouldboards and a hand screw adjuster to give a flat or pointed top ridge. The shorter beamed R.N.R.L. 2 swing ridging plough suitable for Scotland and Northern England had fixed mouldboards. The R.E.P.R. and R.A.R. ridging ploughs without wheels had cast and steel mouldboards respectively and were specially designed for difficult soil types in parts of Lancashire

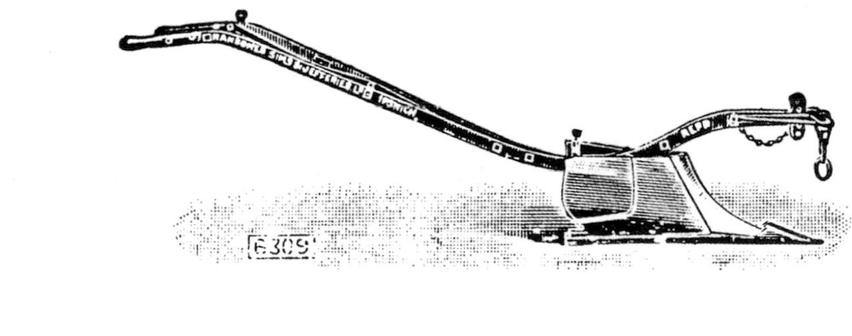

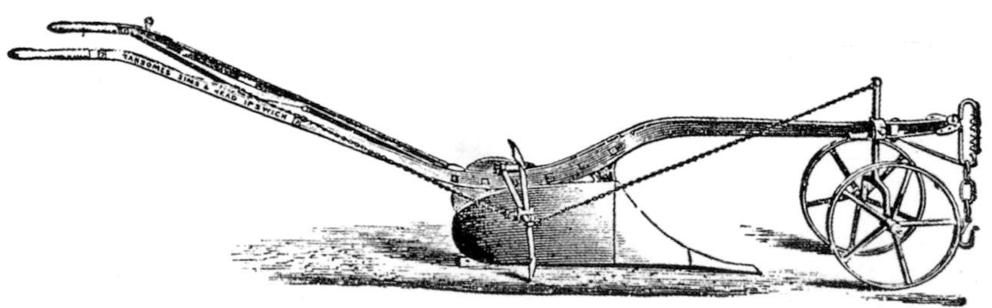

A marker was an optional extra on the improved ridging or moulding plough. Cast-iron mouldboards were standard on the R.E.P.R. ridging plough but the mouldboards of the R.A.R ridging plough (bottom) were made from steel.

and Cheshire. Sales literature explained that the body was designed so that 'light or boggy soil is not mulled or pushed but is worked with ease and freedom and at the same time a perfect ridge or drill is formed.'

The No. 4 and No. 5 horse-drawn ridgers with the main wheels on long expanding axles, a front swivel wheel and steering handle were made in the early 1930s. Three ridges at 24 to 30in spacings were made by the No. 4 ridger which cost £24 15s. The No. 5 made ridges at 24 to 30in spacing when it was fitted with two bodies or 18 to 22in apart when it was used with three.

The No.4 three-row and No. 5 two-row ridgers were made to 'suit users who do not favour a combination implement'.

Potato growers in the late 1930s had the choice of a tractor-drawn No. 4 ridger or either three or four ridging bodies in place of the tines on several models of the Dauntless trailed cultivator. Three ridging bodies could also be used on Ransomes' C25, C26 and C27 rear-mounted toolbars for Farmall, Case, John Deere, Fordson and Allis-Chalmers wheeled tractors (page 46). The toolbar was raised and lowered by a lever-operated chain and ratchet mechanism and a pair of wheels controlled working depth.

The design of horse- and tractor-drawn ridgers hardly changed until the late 1940s when Ransomes introduced the first of a series of three-point linkage mounted toolbars with a range of accessories which included cultivator tines and ridging bodies. The C60, C62 and C66 toolbars were wide enough to make three ridges at a time and up to five bodies could be used on the C68 toolbar. The C73 toolbar for the Fordson Dexta and the C74 for the Major, which superseded the C60, C61 and C62 in 1955, remained in production for nine years. Ridging bodies were available for the C73, C74 and the TCR 1001 toolbar made between 1956 and 1967 for the Dexta. Up to five ridging bodies could be used on the C79 Zed frame toolbar introduced in 1964. Standard bodies were used to make ridges between 18 and 24in wide but the larger bar point body was needed when making ridges from crops grown in rows spaced between 26 to 36in apart. The bar point body was also suitable for use on the C83/86 toolbar.

A steerage fin was adequate when the C79 was hitched to a tractor with a depth-control hydraulic system, and optional depth wheels were available when it was used on tractors with lift and drop hydraulic linkage. Ransomes added the C92 deep ridger in 1981. Two large ridging bodies spaced between 5 and 6ft apart on the toolbar made a flat-bottomed trench wide enough to accommodate the volume of stones separated from some soils by a de-stoning machine.

The Johnson Rotoridger was added to Ransomes' product range in 1968.

Ridging with the FR C73 universal toolbar.

In 1968, when Ransomes aquired Johnson's Engineering Ltd of March, they added to their range of potato machinery, the power-driven Rotoridger and the Johnson three-row rotary cultivator-ridger, which cultivated and ridged up ploughed land in a single pass, to the Ransomes range of potato machinery.

Potato Coverers

Small-scale potato growers covered hand-planted potatoes with a hand hoe or used a horse-drawn potato coverer or a ridger to split the ridged soil back over the tubers. Ransomes' catalogue for 1932 observed that 'covering potatoes lightly with a hand hoe considerably accelerates their growth' but it was claimed that covering the tubers with a No. 2 potato coverer would give an approximate gain of two weeks in the growth of the potatoes.

The No. 2 two- or three-row front coverer was used to split back ridges over potatoes planted in rows spaced between 22 and 30in apart. The horses walked on top of the ridges.

The Ransomes front coverer for the E27N Fordson Major had three ridging bodies and two bobbin wheels bolted to the front-mounted C61 toolbar. The bodies were raised and lowered by a system of rods from the rear hydraulic linkage and the bobbin wheels helped the ridging bodies to split the ridges equally on both sides. The bodies were adjustable for width, depth and pitch and could be spaced out on the toolbar to suit various row widths.

Potato Planters

Manufacture of horse-drawn potato planters was recorded at the Orwell Works in 1882 but in more recent times the acquisition of Johnson's Engineering Ltd, provided Ransomes with a ready-made range of planters. The March company was manufacturing their own

The C61 front toolbar for the E27N Fordson Major was discontinued in 1961.

The Ipswich potato planter, made in the 1890s, planted the crop on previously ridged ground.

two-, three- and four-row hand-fed Johnson potato planters and importing Cramer automatic planters and Cramer Gerlsma automatic models for chitted seed. The new product range made life even more complicated for salesmen at their dealerships as Ransomes were also importing two-, three- and four-row Faun hand-fed planters from Norway. The Faun planting units could be taken off the toolbar without difficulty leaving it free for use as a ridger, cultivator or inter-row hoe.

Ransomes Johnson potato machinery for the 1969 growing season included ridgers, planters, a haulm pulveriser, spinners and elevator diggers, as well as four models of harvester, a bulk transporter, conveyors, elevators and grading equipment. The new Ransomes Johnson Chieftain automatic two-row planter was made at March between 1970 and 1975. The Chieftain had vibrating hopper floors that provided an even supply of seed potatoes to endless belts which carried them to a point above furrow openers, where they dropped into the soil and were covered by pairs of angled discs. An eight-speed gearbox controlled the planting rate and a single operator was seated on the planter to keep an eye on the feed mechanism.

The Johnson range of planters was discontinued in the early 1980s when Ransomes were importing the latest Norwegian-built two- and four-row Underhaug Faun mounted and trailed automatic cup feed planters. The trailed Faun planters had a hydraulically tipped hopper, and a flashing light in the tractor cab alerted the driver to faults in the automatic feed mechanism. Following Kverneland's acquisition of Underhaug in 1986 Ransomes had to find an alternative supply of potato machinery and they introduced a range of Hassia planters and harvesters at the 1986 Smithfield Show. The new range of German-built planters included two- and four-row mounted models, a four-row trailed machine and a three-row bed planter.

In the early 1970s Ransomes marketed the Cramer Automatic planter for un-chitted seed (left) and the Faun semi-automatic potato planter for chitted seed.

The Chieftain automatic planter worked at speeds of up to 7mph and the hoppers held enough potatoes to plant at least half an acre.

Potato Lifters and Harvesters

Potato-raising bodies were made for the Newcastle series ploughs in the 1860s and by the turn of the century Ransomes were also making potato-raising ploughs and potato spinners for horse or tractor draught.

Potato spinners were referred to as diggers in Ransomes' literature and a horse-drawn machine with fixed tines that scooped the potatoes sideways from the ridge was being made in 1890. A screen or curtain was often attached to the spinner to deflect the potatoes down to the ground near the lifted row. A fixed curtain was normally used but the spinner could also be supplied with a circular rotating screen driven by a chain from the land wheel.

The No. 12 horse-drawn potato digger, with the choice of feathering tines or cam-operated hanging tines, was made in the early 1900s and the tractor-drawn version had been added by the 1920s. Ransomes' price list for 1930 included the No. 12, No. 21 and No. 28 potato diggers; the No. 21 with a tractor hitch or horse pole cost £24 2s 6d. A cam-operated mechanism on the No. 21 and No. 28 diggers kept the hanging tines vertical at all times, and the catalogue explained that this design improved the lifting action of the tines when they entered the soil and ensured clean delivery without bruising the potatoes. A lever-operated parallel lift arrangement on the No.28 varied digging depth without affecting share pitch, but the lighter No.21 did not have this refinement and a single lever was used to put the share into and out of work.

The same frame, wheels and gearbox were used for the No. 12 digger with fixed tines and the No. 28A with hanging tines made in 1939. The No. 12 was supplied with a horse pole, whippletrees and seat. With the option of a horse pole, rigid drawbar protected by a spring-loaded safety hitch or head wheel and tow chain, the No. 28A could be used with a tractor or horses. A third horse attachment was available for lifting potatoes in heavy conditions.

An improved No. 28C wheel-driven digger appeared in 1946. The parallel share depth mechanism and optional head wheel or a rigid drawbar were retained but a lever to disengage the drive to the tines was a new innovation. The No. 41 potato digger, introduced in 1946, was Ransomes' first power take-off digger. Later designated the PD3, it had the same tine arrangement as the 28C with a hand screw adjuster for share depth and a cord-operated self-lift clutch.

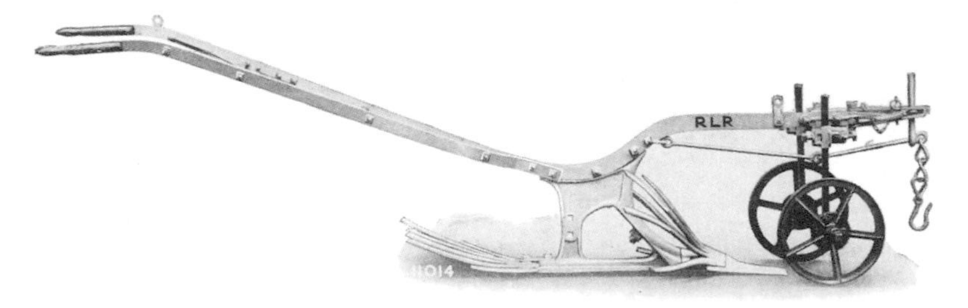

196 A potato-raising body was used on the R.N.R., Newcastle and other horse ploughs.

Ransomes No.12 patent potato digger with hanging tines.

The gearbox and lifting mechanism were retained on the FR mounted potato digger introduced in 1947 and later re-numbered the PD4. Designed for the E27N Fordson Major, it had triple vee-belts to transmit drive from the power shaft to a cast gearbox and the share depth was adjusted with a rubber-tyred wheel. Improved B, C and D variants of the PD4 were made and the PD5, with a gearbox fabricated by Crittalls in Essex, was built during a period when castings were in short supply. The fabricated gearbox was not a success but the FR PD4 potato digger remained in production until the TPD 1008 digger superseded it in 1957. The TPD 1008 was modified several times before it eventually went out of production in 1977.

Johnson single- and two-row mounted and semi-mounted elevator diggers were already popular with potato growers when they were included for the first time in Ransomes' catalogue for 1969. Blue paint replaced the previous orange livery but otherwise they were unchanged. There was a choice of the standard steel rod link elevator or twin belt type of conveyor with rubber-covered links for the two-row mounted digger. To confuse potential buyers, Ransomes were also importing a single-row Faun mounted elevator digger with plain or rubber-covered rod links.

As well as their own bagger and side-elevator potato harvester and J.202 two-stage harvesting and handling system, Johnson's Engineering were also importing Weimar and Faun complete

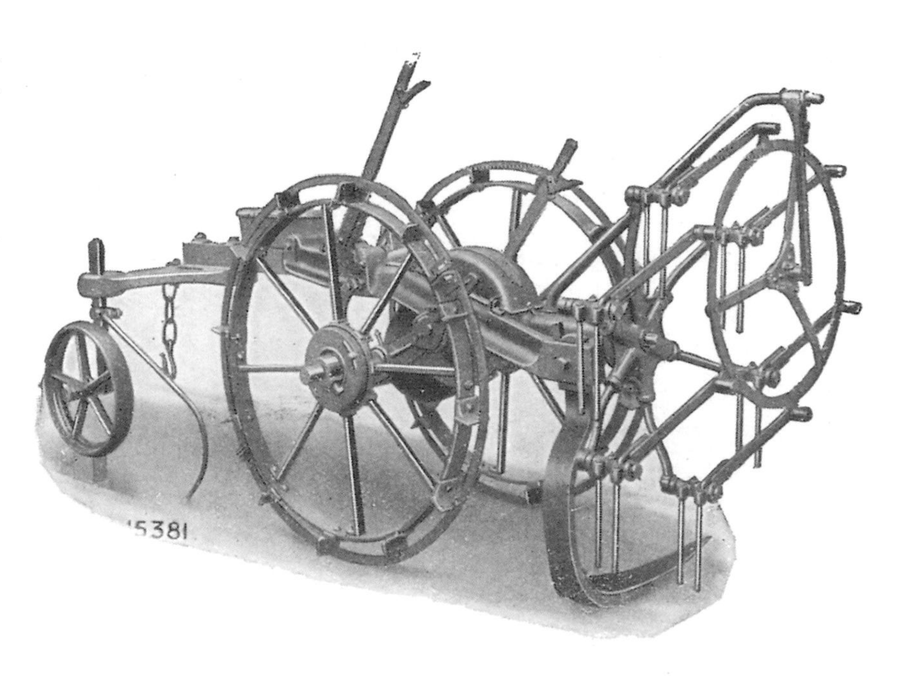

The No. 28A potato digger with hanging tines. The hook at the front was used to clear the tops and make a path for the share arm. The wheel spuds were riveted to the rims, and steel rings were supplied to go over the spuds when taking the machine on the road.

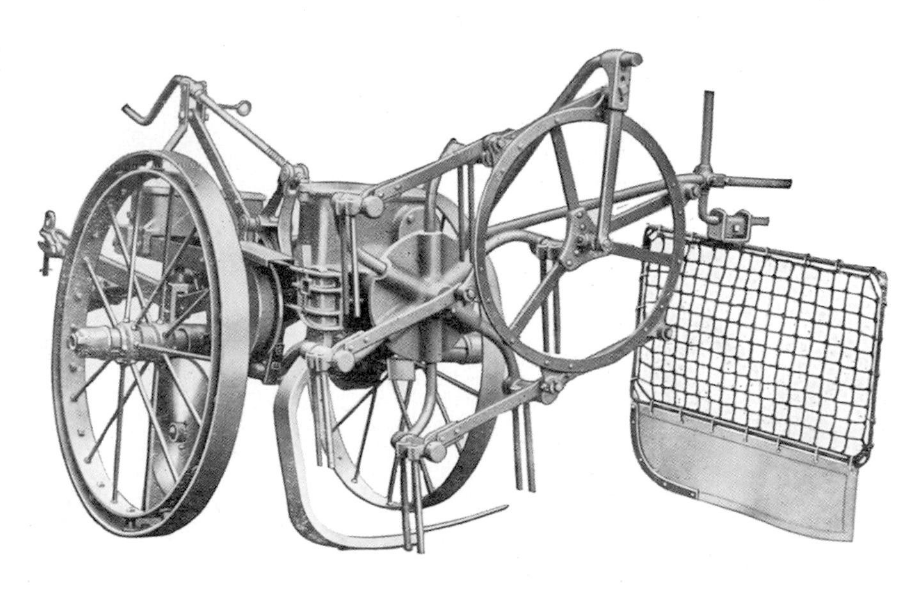

The left-hand wheel of the No. 41 potato digger had an over-tyre to keep the machine level in work.

The new FR mounted potato digger was introduced in 1947.

The last Ransomes Johnson elevator diggers were made in the mid-1980s.

harvesters when they became part of the Ransomes, Sims & Jefferies organisation in 1968. The Ransomes Johnson J.202 two-stage six-row harvesting system employed four tractors with three semi-mounted two-row elevator diggers and a digger loader. At least two other tractors with farm trailers or Johnson bulk transporters were required to haul the crop to the farm potato store where the crop was sorted with a Fulcrop grader. The two central rows of the six were harvested by a straight-through elevator digger and left on the ground behind the machine. Two side-delivery diggers lifted the four outer rows of potatoes, one delivering them to the right and the other to the left, so that the six rows were collected in a single windrow. A digger loader with belt-type conveyor webs collected the crop and side-elevated the potatoes into a trailer. In good conditions it was possible to harvest between fifteen and twenty acres in a day.

Production of the Ransomes Johnson elevator digger continued at March and Ransomes also took over the import of Weimar and Faun potato harvesters. The German-built Weimar harvesters included one with a rotary sorting table and side elevator; another lifted potatoes, stones and clods with two contra-rotating discs which were conveyed on a system of rubber-covered elevators and side-delivered into a trailer. The mixture of potatoes and trash was hauled to the farmyard for sorting and storage or sale. The Faun single-row mounted harvester lifted the crop with a share and rotating disc which removed most of the soil before elevating the crop to a rear-mounted hand-sorting platform. The sorted crop was either side-elevated into a trailer or conveyed to the bagging-off platform attached to the front of the tractor. The bagger model had an output of about 1½ acres in eight hours and required a tractor in

The Ransomes Johnson two-stage harvesting system was designed for farms with at least ninety acres of potatoes grown in flat, free-flowing and friable soil.

the 40 to 50hp bracket. The Superfaun, introduced in 1983, was an improved version of the single-row rear-mounted Faun harvester and, in common with its predecessor, its improved lifting unit was offset to eliminate wheel damage to the crop; it had a larger soil separation sieve, wider elevators and new haulm removal rollers.

Johnson's and Ransomes' engineers joined forces to design the Monarch complete harvester and the first machines were made at March in 1970. Improvements for the 1971 season included variable speed drives, the choice of two sizes of load platform on bagger machines and a new side-delivery model. Described in the sales literature as a low-damage

About two acres could be harvested in an eight-hour day with the side-elevator model of the Faun potato harvester.

An improved and quieter model of the Ransomes Johnson Monarch potato harvester with a 25cwt capacity bagging platform appeared in 1971.

trailed harvester, the single-row Ransomes Sovereign with an output of four acres per day replaced the Monarch in time for the 1979 harvest. The lifting share and elevator web were offset so that the tractor and harvester wheels ran on cleared ground and there was enough space on the Sovereign's sorting platform for eight hand pickers to work under an all-weather canopy. Fifty-one Sovereign harvesters had been built at March when it was discontinued in 1984.

Weimar harvesters were distributed by Bonhill Engineering from the mid-1970s but Ransomes continued to import Faun potato harvesters until Kverneland bought Underhaug in 1986 and Ransomes introduced single- and twin-row Hassia machines with hydraulic rear-wheel steering at the 1986 Smithfield Show.

The acquisition of Johnson's Engineering also gave Ransomes a ready-made collection of potato cleaners, conveyors and elevators driven by an electric motor or small petrol engine. The crop was tipped into a bulk hopper and conveyed to revolving rubber spools that sorted potatoes from stones and soil before they were conveyed into the store. The Minor and Major cleaners sorted the crop at a rate of 30 and 50 tons an hour respectively for bulk storage; when they were fitted with a bagging-off attachment they could deal with up to 8 tons in an hour. Ransomes also inherited the Johnson two-row power take-off driven potato haulm pulveriser with a work rate of eight to fifteen acres a day, Johnson bulk transporters and the Weimar store intake bulk hopper and conveyor.

Sugar Beet

James Smyth at Peasenhall in Suffolk had dominated root and cereal drill market in East Anglia for well over a century before, in a roundabout way, it became part of Ransomes, Sims & Jefferies. Johnson's Engineering bought the Smyth company in 1967 and the remnants of the business were moved to Ipswich when Ransomes acquired Johnson's in 1968. At least 35,000 Smyth drills had been made at Peasenhall and although none were made at Ipswich Ransomes benefited from the sale of spare parts for these drills.

Band sprayers for sugar beet drills were made at

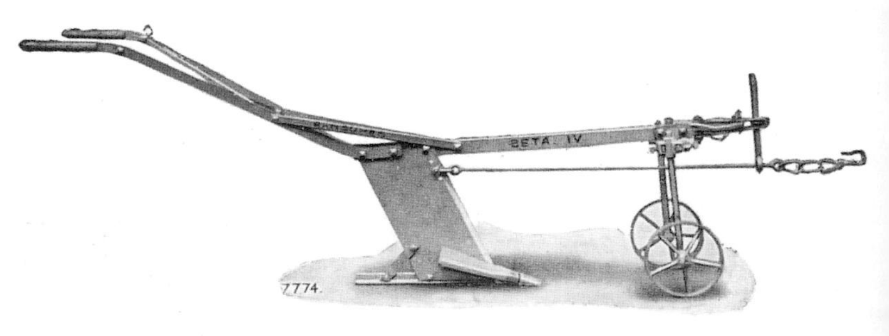

The Beta IV sugar beet lifting plough was described in a 1930 catalogue as 'an extremely efficient implement for raising sugar beet in all conditions'.

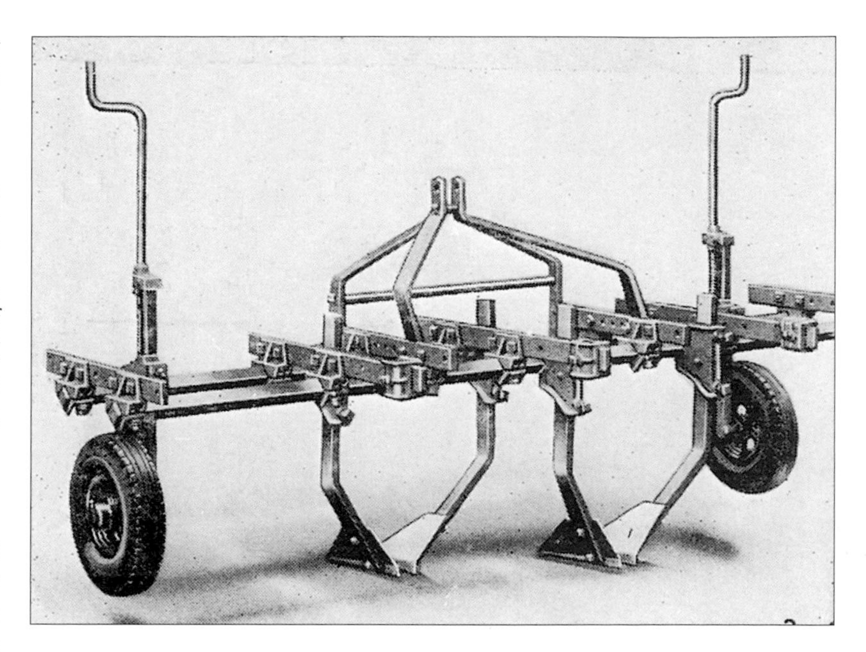

The C60 toolbar fitted with a pair of bow-type sugar beet lifters and pneumatic-tyred depth-control wheels.

Ipswich from the early 1960s for a period of about ten years. Precision seeders were not part of Ransomes' product range until 1986 when, following Kverneland's acquisition of Underhaug, they were appointed UK distributors of Betasem cell wheel and Unisem pneumatic feed precision seeders made by Hassia in Germany.

Sugar Beet Lifters

The horse-drawn Beta sugar beet lifting plough was introduced in the early 1920s and the improved Beta IV with an optional subsoiler leg was included in the 1931 catalogue at £4 10s. Sugar beet lifting shares were also made for some models of two-horse Ransomes ploughs. The Wholroot beet lifter made by J & F Howard was similar to the Beta and this implement was added to Ransomes' catalogue when they bought the Bedford business in 1932.

Sugar beet growers who bought a Ransomes C60 toolbar for the E27N Major were able to use it with cultivator tines, hoe blades and beet lifting shares and in this way save money by not having to buy a separate implement for each of these field operations. It took an hour or two to change the toolbar attachments but there always seemed to be enough time to do it in those days. A similar range of tines, hoe blades and shares were made for the FR C72, C73 and C74 toolbars in the 1950s and for the TCR 1001 introduced in 1963.

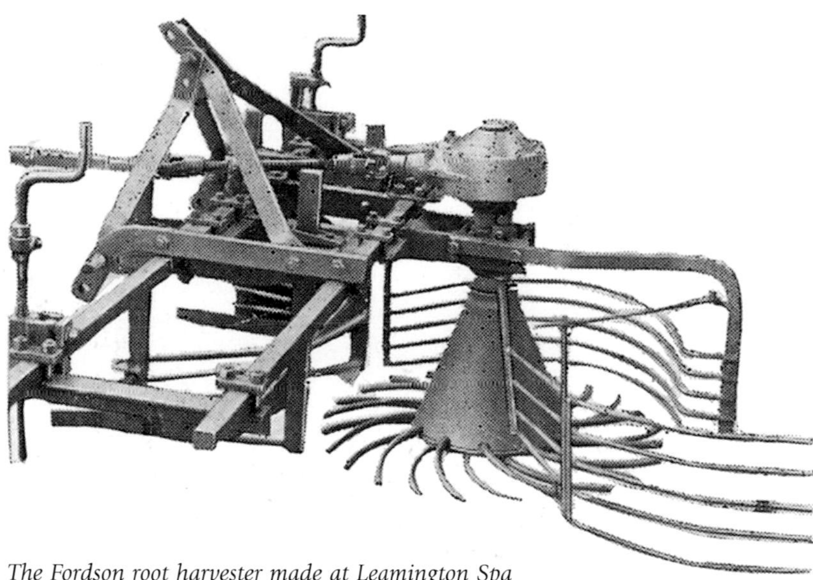

The Fordson root harvester made at Leamington Spa was also sold as an attachment for Fordson and FR toolbars.

The Fordson two-row lifter-cleaner, launched at the 1949 Royal Agricultural Show, was made by the Ford Motor Co. at Leamington Spa until 1955 after which all FR implements were produced at Ipswich. Parts from Ford car rear axles were used in the gearbox on the power take-off driven root spinner that lifted, cleaned and windrowed two rows of previously topped beet and then added two more rows of beet on the return run. It was reported in the agricultural press that by changing the spinner and share the lifter-cleaner could also be used for potatoes and that a conversion kit would be available in time for the 1950 harvest.

Farmer and engineer William Catchpole, who

The Powerbeet was designed to permit the removal of the tractor unit at the end of the sugar beet harvesting season but this was rarely done.

founded the Catchpole Engineering Company at Stanton near Bury St. Edmunds, made a complete sugar beet harvester in the mid-1930s. The single-row, power take-off driven harvester was good enough to be awarded an R.A.S.E silver medal in 1939. The side-elevator Cadet, first seen at the 1954 National Sugar Beet Harvester trials, was still being made in 1967 when Catchpole Engineering launched the 33 tanker harvester.

The single row self-propelled Powerbeet was at a late stage of development when Ransomes acquired the Catchpole business in 1968 and the first Powerbeet harvesters were sold with Catchpole's familiar orange paintwork. The harvester was built around a Ford 4000 or a Massey Ferguson 165 tractor; it weighed less than six tons and the 2½ ton capacity tank was emptied in under a minute. The offset topping unit and lifting wheels ran in line so that the tractor wheels did not need to run between the rows of beet. The Powerbeet had its own rear wheels, and legs were provided to support the front of the machine when the power unit was removed. As Ransomes provided a special drawbar it was still possible to move the harvester around the farmyard. Production of the Super Cadet, Catchpole 33 tanker and Powerbeet harvesters was transferred from Stanton to March in 1971 when they were painted blue and marketed under the Ransomes name. The last Super Cadet made at March was eventually sold in 1974; the Powerbeet, also made at March, was discontinued in 1979.

Ransomes introduced the 33A tanker with an angled trash-removal elevator bridging the tank in 1975. The beet tumbled down the trash elevator into the tank while stray tops and other trash were discharged at the rear. Further refinements were made when the 33B tanker superseded the 33A tanker in 1976. The most obvious changes were an improved trailing topper unit mounted on a parallel linkage with the top removal flails positioned immediately behind the feeler-wheel-adjustable land wheels for alternative row widths. Ransomes sold the last 33B tanker made at Nacton in 1982.

An output of one and a half acres an hour was claimed for the Ransomes Hunter and the 4½ ton tank was large enough to hold the crop until the harvester reached the headland.

The Super Cadet had been discontinued when the two-row self-propelled Hunter beet combine, designed by Ransomes' engineers, made its debut at the 1974 Smithfield Show. A number of components on the Hunter were borrowed from the Cavalier combine harvester, and other features included hydraulically variable speed control in each of the three forward gears and reverse, variable speed tank-discharge elevator and hydrostatic steering. Electronic visual and audible warning units in the cab warned the driver of problems with the beet-conveying systems. A few Hunters were still harvesting beet in the year 2000 and one machine is preserved in the London Science Museum's farm machinery collection.

Barry Mortimer took one of the new Hunters with a Ransomes demonstration team on a tour of four countries on the mainland of Europe. The harvester worked well enough but moving it around on a low loader was another matter. Low and narrow bridges, weight limits, restrictions on weekend travel with heavy goods vehicles and even low overhead signs on the German autobahns were a few of Barry's problems. Then there was the matter of having the correct documents at border posts. In order to get the necessary paperwork Barry had driven over six thousand miles by the end of the tour, which was twice as far as demonstrator Adrian Ward had had to drive the lorry.

Big changes were taking place in sugar beet harvester design and although the 33B and Hunter were still in production, Ransomes looked to the small Danish town of Tim for their next generation of beet harvesters. The first two-row trailed TIM harvesters with a work rate of about six acres a day were imported in 1980 and three-row machines able to lift up to nine acres in a day were added in 1983. Both models were equipped with hydraulically operated

Ransomes imported TIM sugar beet harvesters between 1980 and 1987.

self-steering systems and a 4½ ton hydraulically tipped tank. A new three-row tanker with scalper toppers and elevator to unload the beet into a trailer while on the move had, depending on conditions, an output of up to fifteen acres a day. Ransomes also marketed TIM fodder beet harvesters in the mid-1980s.

Agrolux took over the import of TIM harvesters when they acquired the Ransomes farm machinery division in 1987. The Danish company established a business in the UK in 1991 and marketed harvesters from their Suffolk depot until 1999 when they became part of the Kongskilde group.

Chapter 9

Drills, Mowers and other Farm Machines

Ransomes of Ipswich were best known for their ploughs and field implements, grass-cutting machinery, steam engines and thrashing machines, electric trucks and trolley buses and more recently combine harvesters and root crop machinery. However, during their 200-year history, many other types of machine were made at the Orwell and Nacton Works. The list is a long one which includes celery moulders and lifters, corn mills, food-processing machinery, haymakers, printing machines, saw benches and trailers.

Drills and Fertiliser Distributors

Cultivator drills are not a modern invention. Grain and small seed boxes to broadcast the seed while cultivating the ground were advertised by Ransomes in the early 1900s. The corn and grass seed box, suitable for use with the Ipswich Steel and other Ransomes horse-drawn cultivators, kept the horseman fully occupied. He was required to raise and lower the cultivator tines, keep an eye on the seeding mechanisms and at the same time keep his horses on a straight course across the field. Hand-propelled seeder units and Hornsby grain drills were included in Ransomes' farm implement catalogues in the 1920s and 1930s during their association with the Lincolnshire company.

Smyth cup-feed grain and root drills had been made at Peasenhall in Suffolk for about 170 years when Johnson's Engineering bought the business in 1967 and transferred production to March in Cambridgeshire. The cup-feed mechanism had always been at the back of the drill under the watchful eye of the horseman until 1962 when the hopper was reversed to put the feed cups in full view of the tractor driver. The drill was still made in this way when Ransomes acquired Johnson's Engineering. Although it was the end of the line for Smyth and for Johnson Smyth cup feed drills there was still plenty of money to be made selling spare parts which were kept in stock at Nacton for the next ten years or so.

Ransomes were making corn and grass seed sowers in the early 1900s.

Automatic markers and following harrows were standard equipment on Ransomes Nordsten Lift-o-matic grain and combine drills.

Ransomes brought three Nordsten drills into the country from Denmark in the spring of 1966 for test and evaluation during the autumn drilling season, and a young Bruce Dawson travelled three thousand miles with a Land Rover and trailer to introduce the drills to dealers. A marketing agreement with Nordsten was announced in 1967 when six models of Lift-o-matic grain drill and two universal combine drills were exhibited at the Royal Show. Sowing widths ranged from 2.5 to 6m; studded nylon feed rollers were driven through a sixty-speed gearbox and the Suffolk-type coulters were spaced at 4in intervals. Corrosion-resistant nylon feed rollers were also used for the fertiliser feed mechanism on the combine drills. A Nordsten 47 coulter grass seed drill with a 3.9m sowing width and studded-roller feed was added to the range at the 1968 Royal Highland Show. The wider than usual Suffolk coulters were spaced 3¼in apart and deflectors at the base of the coulters spread the grass seed in bands approximately 1¼in wide.

Lift-o-matic drills automatically engaged the feed mechanism and lowered the appropriate marker when the driving wheels made contact with the ground but the coulters did not enter the soil until the drill moved forward. Drive was disengaged when the drill was raised at the headland.

New 2.5 and 3m Combi-matic combine drills also with narrow row spacings were added to the Nordsten range in 1971. Separate fertiliser coulters, running between each pair of seed coulters, placed fertiliser in bands about 1½in below and 2½in to one side of the seed. Sixty seed and fertiliser application rates were available through a quick-change gearbox and independent depth adjustment was provided for the fertiliser coulters.

A new 3m Nordsten cultivator drill for heavy clay soils was announced at the 1976 Royal Show. Twenty-one seed tubes were attached to spring tines at 5½in spacing and sowing rate was varied through a sixty-speed gearbox. A similar 4m Nordsten cultivator drill with twenty-seven narrow-spaced coulters was added

This 6m Exact-o-matic fertiliser distributor had a side-filling device which filled the hopper in about six minutes.

in the following year.

Eleven models of Lift-o-matic drill were listed in Ransomes' Nordsten sales literature for 1982. There were six grain drills with twenty-one to forty-nine coulters and 2.5 to 6.0m sowing widths as well as 2.5 and 3m universal models and a 3.9m grass seed drill. Two cultivator drills with twenty-one or twenty-seven coulters and 3 or 4m sowing widths completed the range. Optional equipment included a transport carriage, hopper extension and a tram lining kit.

Nordsten Exact-o-matic mounted fertiliser distributors were included in Ransomes' price list for the first time in 1972. There were four drill widths from 2.8 to 5.4m with nylon-studded roller feed. Three new 3.8, 4 and 6m, full-width Exact-o-matic fertiliser distributors replaced the earlier models at the 1974 Smithfield Show. Mounted and trailed models were made and all three were end-towed for transport purposes. A side-filling mechanism with an auger placed at the top of the hopper to convey the fertiliser across its full width was optional. The auger, driven by a hydraulic motor, was automatically stopped when the hopper was full.

Ransomes added the Air-o-matic pneumatic fertiliser distributor to their range of Nordsten machines in the early 1980s. The mounted distributor with a 1,000 litre capacity hopper had twenty fertiliser outlets spaced across a 20m boom and the metering unit was driven by a rubber-tyred land wheel. A trailed version of the 12m Air-o-matic distributor and a Mk II mounted model, with a new two-point quick hitch and lower filling height among a number of other improvements, were announced at the 1983 Smithfield Show.

Howard Farmhand, part of the Scandinavian Thrige Agro group, based at Wymondham in Norfolk, marketed Nordsten drills and fertiliser distributors in the UK from 1987, when Ransomes' farm machinery division was sold to Agrolux. Kongskilde became Nordsten drill importers for the UK in 2001.

Mowers and Hay Machinery

The first Ransomes FR mounted field mowers made by Sankeys in 1958 had a 5ft cutter bar and cost £99 10s. The 5ft cut and a new 6ft cut mower, now made at Ipswich and designated the TM1021 rear mounted mower for the Fordson Dexta and Power Major, were included in the Ransomes price list for 1959. The price of the 5ft mower was unchanged but the new 6ft model was an extra £4 10s. Features of the TM1021 included a tubular frame with an exclusive spring suspension system and

The TM1021A field mower was approved for use with the Fordson Major and Dexta.

break-back safety release for the cutter bar.

The suffix letters A, B, C and D used with the TM1021 model number indicated the geometry of three-point linkage and relative position of the power take-off input shaft for different makes of tractor. The 5 and 6ft cut TM1021A was approved for use with Fordson Major and Dexta tractors and the TM1021B, with additional guards to meet safety regulations, was approved for Ford 1000 series tractors built at Basildon. The 5 and 6ft cut TM1021B cost £121 and £126 respectively. Sales literature for them explained that the unique spring suspension system was the prime reason for their smooth running, few repairs and long life. The literature went on to claim that in the opinion of thousands of users the TM1021 was 'the smoothest-running, sweetest-sounding mower'.

The TM1021C, added in the early 1960s, had a power shaft extension and rear drawbar to allow mowing and conditioning to be carried out at the same time. A stronger frame and guards complying with the latest farm safety regulations were the more obvious improvements on the TM1021D mower which replaced the previous model in 1970. The prices of the 5 and 6ft cut mowers had risen to £175 and £185. As most farmers were buying one of the more efficient disc or drum mowers by this time, Ransomes discontinued field mower production in the late 1970s.

The FR TG3007 rotary slasher announced at the 1959 Smithfield Show with a £160 price tag was originally designed for the export market to cut and pulverise scrub and saplings up to 3in thick. It was equally successful in the UK and was used to shred kale stalks and other crop residues, cut long grass, thistles, bracken, etc., and slash heavy undergrowth to make fire breaks in forestry plantations. Two heavy-duty free-swinging blades attached to a central power take-off driven hub cut a

The N.I.A.E. testing station in Kenya found the FR TG3007 rotary slasher suitable for shredding the tough fibrous roots of the pineapple crop.

The Jarmain patent swath turner.

66in wide swath while positive height control between 2 and 16in was provided by the cushion-tyred rear wheel. Optional equipment included shredding chains that could be used in place of the blades and a rear skirt panel to confine the cut material for finer shredding.

With a full set of guards and standard power take-off drive unit, the rotary slasher cost £363.50 in 1972. A safety slip clutch for the power shaft added another £33.50. The rotary slasher was discontinued in the mid 1980s.

Horse-drawn hay rakes and haymaking machines were made at the Orwell Works for the best part of one hundred years. Hay rakes were already in the Ransomes catalogues by 1845; the first Jarmain patent swath turner was made in 1892 and, with the exception of a period between 1914 and 1918, hay rakes were built at Ipswich until the start of the Second World War. The one-horse Jarmain swath turner had two sets of rotating paddles driven by the wheels, which moved two swaths sideways on to dry ground.

In 1890, when nearly thirty-five thousand Star hay rakes had been sold, a catalogue suggested that these rakes were 'so widely known and so generally appreciated that any lengthened description of them was unnecessary'. There were four sizes of rake spanning 7 to 10ft in width and in three capacities with varying sizes of 'T' section steel teeth and height of the wheels. Prices ranged from £8 to £12 and a seat and hand lever were an extra 15s.

The Ransomes Through Axle horse rake was designed to meet a demand for a cheaper rake. Fitted with a solid full-width axle, the rake had a seat and hand lever included in the price. A second lever was provided for use when walking behind and the leverage was so light 'that a strong lad could work it without difficulty'. The catalogue also sang the praises of the new self-acting horse rake awarded the first prize gold medal at the International Agricultural Exhibition held at Amsterdam in 1884. The rakes were 'entirely self-acting in delivering the load so that a boy can work them'. All he had to do was to put his foot on a treadle which caused ratchets on the road wheels to lift the tines to deposit the load and then drop the tines back into work.

Three sizes of Star and Ariel horse rakes and two models of the Ipswich self-acting rake were still in production in the mid-1930s. The Star was much the same as the original 1840s version; the Ariel and Ipswich were made in self-acting or manual format and the Ariel was also available as a combined manual and self-acting hay rake.

Ransomes also made haymakers which, in modern terms, were overshot or undershot tedders. The Ipswich haymaker, also awarded the first prize gold medal for haymakers at the Amsterdam exhibition in 1884, had both overhead (overshot) and backward (undershot) tine actions. Made in two sizes, with the larger one taking two mower swaths, the Ipswich haymaker was 'light in draught and confidently recommended as

Various models of Ransomes Star horse rake were made for more than ninety years.

The attractive cover from a late 1800s hay machinery catalogue.

thoroughly strong, simple and durable'. The Star haymaker was a two-speed machine with only a backward action that cost £14; farmers requiring a cheaper machine could save £3 10s by buying a single-speed back-action model. Ransomes' hay machinery was popular with farmers, both at home and abroad, and by 1900 they were able to advertise that fifty thousand of their hay rakes and haymakers were in use.

Farm Transport

Farm trailers made by F.W. Pettitt Ltd at Spalding appeared in Ransomes' catalogues in the late 1950s and were sold in the company colours for about thirty years.

The FR TT1017, a 3 ton hydraulic tipper, and the TT1018 non-tipper, both of which had wheels at the rear, a 10 x 6ft wooden floor and wooden sides, were marketed by Ransomes and approved for use with Fordson Major and Dexta tractors. The TT1017A was the same trailer but had a dual ring and jaw hitch and provision was made for moving the axle forward when using the jaw hitch. The TT1026, a 3½ ton tipper, also with a wooden floor and sides and 10in diameter drum brakes, was added in the early 1960s. Optional equipment for both trailers included road springs and lights, grain sides, double-height silage sides and tailgate, timber harvest ladders and rear floor extension.

Eight new models of two- and four-wheel TT tippers, two sizes of TU three-way tippers and the TN2314 two-wheel non-tipper with capacities of 3 to 8 tons were launched at the 1967 Smithfield Show. The TT4600, 4700 and 4800 tippers had tandem rear wheels and twin rams. Hydraulic trailer brakes were optional and the usual selection of harvest ladders, grain sides and other accessories was available. Prices started at £175 for the TT2300 3

The TT2300 3 ton tipping trailer with 18in high timber sides and a wooden floor cost £174 in 1968.

Ransomes' 4 and 6 ton scissor high-lift tipping trailers appeared in the late 1970s.

ton tipper and the largest TT4800 tandem-wheeled 8 ton tipper was £435 5s. A 10 ton tandem four-wheeler introduced in 1968 completed a twelve-model range of Ransomes tipping trailers. With the exception of the 3½ ton TT2310, the other trailers in the range were still in production when the 12 ton TT4126 tipper was added in 1976.

Two sizes of bulk potato transporter with capacities of 4 and 8 tons were added in 1968 when Ransomes acquired Johnson's Engineering. They were filled directly from the harvester and unloaded into the store intake hopper by an endless rubber-covered floor conveyor driven by a hydraulic motor, petrol engine or electric motor.

The FR TG1020 tipping transport box, introduced in 1958 and made for nine years, carried half a ton. The load platform was big enough to carry eight 12-gallon milk churns.

A 6nhp engine was powerful enough to drive a Ransomes double steam flour mill with 36in diameter stones; twice as much power was needed for a double mill with four 54in diameter stones.

Other Products

As mentioned (page 8), Ransomes produced vast quantities of railway sleeper chairs, trenails and other components including iron work for bridge and station construction during the railway building boom that started in the mid-1830s. This work continued until 1869 when it was decided to concentrate on the agricultural business and establish a new firm called Ransomes & Rapier to handle work for the railways.

A widened range of agricultural machines included steam-powered water pumps, saw benches and thrashing machines, chaff cutters, corn dressers and corn mills for grinding flour. Wooden-framed corn mills driven by a set of horse works dated back to the 1830s. The arrival of steam power saw various sizes of fixed corn mills added to Ransomes range of portable corn mills and by the 1880s Ransomes were making batches of two or three iron-framed single and double mills for grinding wheat, maize and other grains into flour and animal feed. Single mills were made with various sizes of French Burr or Derbyshire stones from 30 to 50in diameter and any number of up to six mills could be placed side by side and driven by a portable or fixed steam engine. The stones had to be re-grooved with a hand tool from time to time, and a crane arm attached to the mill framework for lifting the top stone was an optional extra. There was a thriving export trade for steam flour mills, and, in countries where the alternative wooden frame and stones could be obtained locally, customers were able to buy just the iron work and gearing and thus reduce the cost of transport.

Arthur Biddell's scarifier was not the Playford farmer's only invention. He also held patents for a root cutter, an oat mill and a bean cutter made at the Orwell Works in the mid-1800s. Other Ransomes food-preparation machinery made at the time included roller mills, oil-cake breakers, corn dressers and the horse gear, or horse works, used to drive this equipment through a system of belts, pulleys and shafts. In 1827 Ransomes, Sims & Head sold their own and Biddell's patent food-preparation equipment designs to Reuben Hunt at Earls Colne in Essex.

Horse-drawn celery moulders and lifters were introduced in the 1930s to reduce the amount of hand labour required to grow this crop. The moulder was used after throwing up soil on both sides of the row with a ridging plough. Steered by one man and pulled by two horses, it raised the soil to the top of the ridge and at the same time firmed it against the plants. The lifter, introduced to 'facilitate the arduous work of taking up celery roots', consisted of a steel frame on four wheels with small left- and right-handed plough bodies to turn soil away. A steel cutting blade ran under the fibrous roots to cut them, after which, as the catalogue explained, the crop could be easily lifted and laid out ready for bundling.

Various company acquisitions in the late 1960s added sugar beet harvesters, potato planters, harvesters and handling equipment described in Chapter 8. Other Catchpole Engineering contributions included the Blackwelder root gapper, Deben and Anglia land levellers, Suffolk land grader, cage wheels and a rear-mounted hydraulic fork lift. This equipment, together with the Super Cadet, the 33 tanker and Powerbeet sugar beet harvesters, was included in a separate Ransomes Catchpole farm machinery price list for 1970. The sugar beet harvesters and forklift remained in production at Stanton but the other lines were withdrawn when existing stock had been sold. The FR hydraulic mounted forklift with a tilt control and a maximum lift of 15cwt to a height of 7ft 3in was discontinued in 1972.

The ban on straw burning opened up a new sales opportunity and Ransomes were one of several companies to add a straw chopper to their product range. Introduced in 1985, the new tractor-mounted chopper required a 65 to 100hp tractor to drive the 2.3m wide rotor at 2,000 rpm. With different shaped flails, the straw-chopper's use could be widened to include chopping light scrub and crop residues such as kale and maize stalks as well as for topping grassland.

The Ransomes celery moulder was, according to a 1932 catalogue, intended 'to do away with the manual labour involved in the moulding up of celery'.

Chapter 10

Electric Vehicles

Electric trucks and lawn mowers, trolleybuses, dynamos and electric motors were made at various times during the 200-year history of the Orwell and Nacton works. When the steam era was drawing to its close, Ransomes looked for new products, making their first British battery electric truck in 1918 and building their first trolleybus in 1924. The Ransomes Electra lawn mower, introduced in 1926, was the first British-built mains-electric lawn mower. Electrically driven 'Wobblums' amusement cars and mechanical horses for fairgrounds and seaside piers were introduced in 1927.

Trolleybuses and Trucks

Horse-drawn trams appeared on the streets of Ipswich in 1880 and within five years there were eighteen horses and eight tramcars working four miles of track. Electric trams, which completely replaced the town's horse-drawn vehicles by 1903, remained in service for twenty-three years.

The English Electric Company was making single-deck trolleybuses in the early 1920s, the Ipswich Corporation hiring three of them in 1923. The pay-as-you-enter trolleybuses with twenty-two seats inside and eight open-air seats for smokers at the rear were an instant success and within a year the Corporation had purchased the three hired vehicles. A fourth single-deck trolleybus, this time made by Ransomes, Sims & Jefferies, was added to the fleet in 1924. When the last Ipswich tram was withdrawn from service in 1926 the Corporation had a fleet of thirty trolleybuses half of which had been made by Ransomes and half by Garretts at Leiston.

It was not uncommon to see Ransomes' staff using the overhead wires on Felixstowe Road to test trolleybuses before their delivery to various destinations in the UK or overseas including Malaya, New Zealand, South Africa, Switzerland and India. Georgetown in Penang, part of the Federated Malay Straits, was the first overseas customer. In 1927 three low-chassis type C trolleybuses, with 50 bhp Ransomes electric motors and pneumatic tyres, were delivered to Penang, followed in 1928 by five more. In 1935 Ransomes sent fifty trolleybuses to Cape Town and during the 1930s they sent another twenty-four of various types to Georgetown, some of which were supplied in chassis format with the coachwork added locally.

Four Ransomes double-deck trolleybuses, Nos. 46 to 49, were

An early Ransomes trolleybus.

Ransomes' 48volt battery electric tipper lorry No. 26696, made in about 1914, had solid rubber tyres.

added to the expanding Ipswich Corporation fleet in 1933 and double-deck vehicles remained on the streets of Ipswich until August 1963. The Royal Agricultural Show was held at Ipswich in 1934 on land now occupied by the Chantry housing estate on the outskirts of the town. The overhead power wires were extended along the London Road so that the new double-deck vehicles could take people to the showground.

Battery-electric lorries, some chain-driven and others with an electric motor in each front wheel, were made between 1914 and 1925. The County Borough of Ipswich and several local companies used Ransomes 2½ and 3½ ton battery-electric lorries with wooden-spoked wheels and solid rubber tyres to collect and deliver goods in the Ipswich area.

The Orwell Electric Tower Wagon, based on a 2 ton lorry chassis and made about 1919, had a top speed of 12mph and a range of thirty to forty-five miles with fully charged batteries.

The Borough bought a battery-electric tipper truck operated with a screw mechanism and a second one was used as a dustcart in 1921. The driver had the added luxury of a cab with a solid roof and sides, which was a considerable improvement on the canvas hood of earlier models.

Ransomes Orwell Electric lorry was driven by an electric motor in each wheel. The basic vehicle weighed 1 ton 11 cwt but with 17 cwt of batteries it tipped the scales at 2 ton 8 cwt.

Electric Trucks and Forklifts

Ransomes built and tested a prototype battery-powered electric vehicle in 1915 before introducing Britain's first commercial battery-electric platform truck in 1918. A battery-electric runabout crane was exhibited at the 1921 Motor Show and the first Ransomes four-wheel electric reach trucks with the forks designed to pick up pallets and boxes appeared in 1927. The pallet could be loaded on to a waiting lorry or lowered on to a load-carrying platform on the truck and carried to the required location. A system of cables was used to operate the forks and stow the load on the truck but muscle power was needed to push the truck from place to place.

Fixed- and elevating-load platform electric trucks were introduced in the early 1920s and made at the Orwell Works for the best part of thirty years. Elevating trucks were mainly used to pick up and transport stillages (boxes on short legs to store components for further machining or use on the production line) around factories, and many of them were used for components in the Orwell works. There was a choice of a high, low or narrow fixed-load platform or an elevating-load platform with a 1, 2 and 4 ton payload. The load platform on an elevating truck could be raised 4½in from its minimum height of 11in to a maximum of 15½in. Hand-operated hydraulic gear raised and lowered the platform on 1 ton elevating trucks and a 2hp electric motor did this job on 2 and 4 ton trucks. The electric motor for the 1 ton truck was mounted on a turntable above a single wheel and a dead man's handle was used to steer the vehicle, select one of the four forward and four reverse gears and apply the brake. It had a top speed of about 5mph, and stabiliser wheels on spring-loaded castors were attached to both sides of the main frame.

There were two- and four-wheel steer versions of the 2 and 4 ton trucks. The driver stood on a small driving platform to control the vehicle with a tiller-steering

Ransomes' first electric reach truck made in 1927. The load was picked up on the forks and either lifted on to a lorry or carried on the load platform on the truck.

handle, interlinked with a foot pedal. Both models had four forward and four reverse gears and a top unladen speed of about 5mph. Variants included a tipping hopper truck for the bulk handling of loose materials, a crane truck, ride-on three- and four-wheel tug tractors and a high-lift 2 ton tiering truck. This was an early type of forklift truck with an elevating platform capable of lifting a 2 ton load to a maximum height of 5 or 6ft in less than a minute. Tiering trucks were particularly useful for loading or unloading heavy goods from lorries and railway wagons.

About one thousand two hundred Ransomes 1, 2 and 4 ton battery electric trucks were made between 1939 and 1945 as part of the company's war effort. A batch of 1 ton trucks was sold to the Bristol Aeroplane Co. and a large number of 2 ton elevating trucks and 2½ ton tiering trucks were sent to Russia.

A new 1 ton battery electric Forklift 20 launched in 1947 was followed within a year or so by the 30cwt Forklift 30 and 2 ton capacity Forklift 40. Specifications included rear-wheel steering, hydraulic lift and tilt mechanisms, parking brake and a top speed of 6mph. All three trucks had a towing eye incorporated in the rear ballast weight and optional equipment included solid rubber or pneumatic tyres and various masts with maximum lift heights of 8 to 14ft. Foot pedals controlled drive to the front wheels and operated the independent front-wheel brakes.

The driver controlled Ransomes fixed and elevating platform battery electric trucks while standing on a small platform. The crane truck version carried 1½ tons on its load platform and could lift up to half a ton with the crane operated by a separate 1.85hp electric motor.

It took twelve hours to recharge the batteries on a 2 ton Ransomes electric truck and when fully loaded it had a top speed of 7mph.

The TNU electric tiering truck, which lifted 2 tons to a height of 5 to 6ft in under a minute, was the forerunner of Ransomes' electric forklift trucks.

The driving unit on the Type TE 20/30cwt electric tractor was spring mounted on a turntable that gave 'an amazingly small turning circle'.

A crane attachment, fork extensions, lights, battery charger and canopy to protect the driver were optional extras for Ransomes Forklift 20, 30 and 40 battery electric trucks.

In 1959 a marketing agreement with truck manufacturer Hyster of Portland gave the American company worldwide selling rights for Ransomes electric-powered trucks, while Ransomes sold Hyster petrol and diesel trucks in the UK. This arrangement continued until April 1973, when the electric truck division announced, 'We're back in the driving seat' and customers were able to buy their electric trucks and tug tractors direct from Ransomes.

A 4 ton electric fork truck and two Four-Way travel reach trucks with lifting capacities of 2,100 and 2,500lb were launched in 1973. As the name suggests, the Four-Way travel reach truck was designed to travel forwards, backwards and sideways to facilitate handling long loads in materials stores and similar confined areas. The trucks had out-rigger wheels to maintain stability in all directions. A single control lever activated the Four-Way drive facility and an indicator on the instrument panel displayed the selected direction of travel.

Several models of the L-Range fork truck, including the 2,000 and 4,000lb capacity L20A and L40A, were launched in the mid-1960s. The L-Range was gradually extended and by 1978 there were ten models, from the 2,000lb capacity L20M to the top of the range L80M with a maximum lift of 8,000lb. There was a choice of a standard or high-lift mast for the cushion-tyred trucks with a 48 volt electric motor, single-pedal Solectronic control and hydraulic mast motor, all designed and built by Ransomes. Low profile Lo-Range versions of the L-Range trucks with 1,500 to 8,000lb capacities for working in low buildings and ship cargo holds were also built from the early 1970s.

Ransomes were making three- and four-wheeled electric tractors with solid rubber tyres in the 1940s and 1950s. The three-wheeled tractor pulled 30cwt at 6mph and was controlled with a single handle also used to steer and apply the brake. The larger and more sophisticated T3 electric tractor could pull up to 6 tons at about 3mph; it had steering wheel, brake pedals, forward and reverse speed controller and a cushioned seat. Lift trucks, reach trucks and T-Range battery electric industrial tow tractors were marketed under the Hyster-Ransomes agreement between 1959 and 1973.

The Orwell Works Fire Brigade

The history of the company fire brigade dates back to the early 1840s when there were two manual fire engines at Old Foundry Road. Within a few years the voluntary fire brigade and their equipment had moved to the Orwell Works. A small manual pump was stationed at the Parkside timber yard and a larger engine, made by Ransomes when they were at Old Foundry Road, was taken to the Orwell Works when they moved there in 1849. A team of thirty-two men operated the large pump but in an emergency twenty were considered sufficient. There was a gentleman's agreement that the Orwell Works Fire Brigade would assist the Ipswich Borough Brigade and they were called out on several occasions to tackle serious fires.

A large tank at the Parkside was filled with water from a nearby pond. When the water was pumped at 82 gallons a minute, there was enough in the tank to keep the pump going for at least an hour and a half. After that it was necessary to move the pump to the dockside or the Parkside pond. Ransomes' fire-fighting equipment was improved in 1870 when at least seventy-two fire buckets were placed in various sections of the works, and chemical extinguishers were introduced nine years later. It was a practice within the Orwell Works Fire Brigade to provide the firemen with beer. A boy carried it to the scene and for a fire lasting two hours at the works each man was provided with two pints. Hydrants and hoses were installed in the various departments during 1891 and 1892 and from that time the manual engine was only used away from the works when assisting the Borough Fire Brigade. The Orwell Brigade assisted the borough for the last time in 1923 when they attended at Burton & Son who were provision merchants in College Street. With more up-to-date fire fighting equipment dispersed around the works, the manual pump, then seventy-five years old, was given in 1925 to the Suffolk village of Fressingfield.

The threat of enemy action in 1939 resulted in the purchase of a motor trailer pump and a 10cwt van to carry

the hoses, which were stationed in the Parkside fire engine house. Stirrup pumps and buckets were placed around the works and a second trailer pump, added in 1940, was stationed in the lawn mower works. The Orwell Works was under Admiralty Control during the war period and the company was loaned an additional 350 / 500 gallons per minute Scammell pump in 1941 followed by a smaller trailer pump in 1942. The National Fire Service also loaned a small trailer pump to Ransomes, which was stationed in the No.2 Plough Works and manned by N.F.S. men employed by the company. In total there were some two hundred employees, all volunteers, in the Orwell Works Brigade who gave their spare time for training and on several occasions dealt with fires outside the works area.

With the war at an end the number of men involved was reduced to peacetime levels and those who remained took part in various fire drill competitions with considerable success over a number of years. One plough works employee and member of the fire brigade during the Second World War left Ransomes in 1946 to pursue interests in the world of entertainment. Percy Edwards became a household name for his bird and animal impersonations on the radio in a second career that spanned forty years.

Development of the Nacton Works during the 1950s and 60s with its new buildings made it easier to introduce modern fire-fighting equipment. The main purpose of the Fire Brigade with its Dennis pump and two trailer pumps was to promote fire prevention, and if an outbreak did occur to deal with it quickly. The largely unused part of Nacton heath within the factory area presented a new fire hazard and there was a major gorse and scrub fire there during the hot summer of 1976. The Ipswich Fire Brigade contained the blaze along the boundary with Cranes factory premises while the Ransomes Brigade used their Commer fire tender and a light trailer pump to prevent the blaze spreading to the nearby railway line. The Ransomes team were occupied on and off for about three days in damping down the matted vegetation. The Commer tender remained in service at Nacton until it was sold in the late 1980s.

The Orwell Works fire brigade assembled during the Second World War.

When sales came back under Ransomes' control, they sold a considerable number of T12 electric tractors to British Airways for baggage handling, and in 1979 they delivered their one-hundredth T12A electric tractor to Heathrow. Also in 1979, they launched standard and high-speed models of the T25M tow tractor. The standard model could haul 16 tons at 5½ mph, and the high-speed T25M had an unladen speed of 15mph and pulled an 8 ton load at a top speed 10mph. Drivers sat in the open air on the early T-Range tractors but in later years they had the luxury of a sound-proofed cab with full spring suspension which, according to sales literature, made them 'a joy to drive'.

The electric truck division was still building a wide range of forklift trucks, battery electric tractors and electric motors in the mid-1970s. A Company Report noted that although the forklift truck business was very competitive Ransomes were in a good position to produce trucks to special specifications which could not be met by large-scale manufacturers. Ransomes' forklift trucks at that time included ten L-Range models on solid cushion tyres and a lift capacity of 1,000 to 8,000lb, as well as four E-Range models with large-capacity pneumatic tyres suitable for use on uneven surfaces and able to handle loads of 3,000 to 8,000lb.

The new R series of electric reach trucks announced in 1974 was designed to operate in narrow aisles between warehouse racking. The R35C and R45C reach trucks were suitable for indoor and outdoor applications. They could be used either for loading warehouse shelves or else for high-lift loading packing cases and other items on to flat-deck lorries. The R35D and R45D with added hydraulic capacity and increased lift heights replaced the R35C and R45C in 1979.

The E80A heavy-duty pneumatic-tyred lift truck was introduced in 1977 and superseded two years later by the E80M. It was a versatile vehicle designed to work inside and outside in wet conditions, to run over sunken rail tracks on docksides and to operate in a variety of locations where trucks with internal combustion engines were considered environmentally unfriendly.

A limited number of battery-electric trucks made at Nacton in 1980 were assembled and prepared for sale by the Hamech group who held exclusive selling rights for all Ransomes' electric trucks and tractors. Following the hand-over of the Electric Truck Division to Hamech later in the year, the vacated space was used for crop-sprayer production.

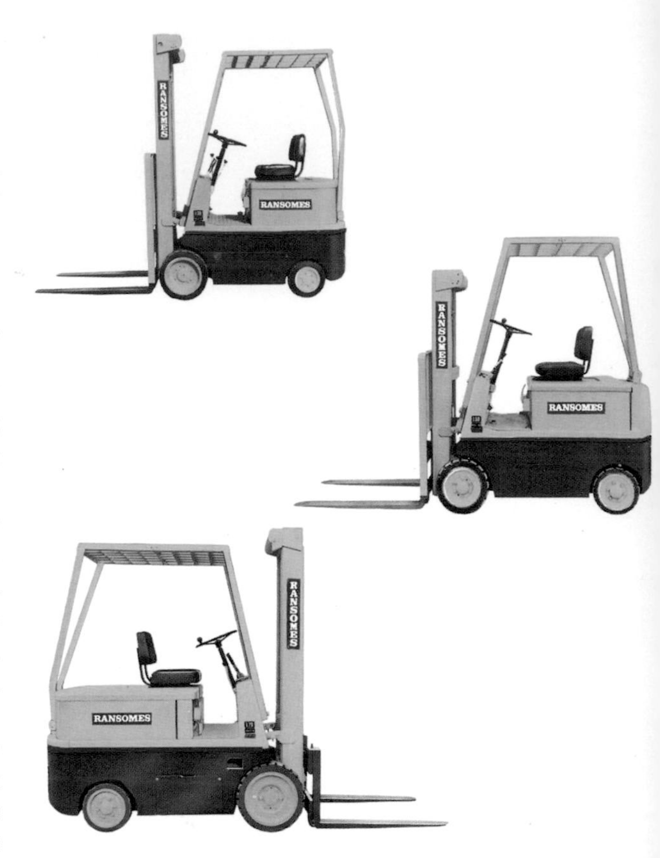

Ransomes built L-Range fork trucks with lift capacities of 2,000 to 8,000lb between the mid 1960s and the early 1980s.

1,500 to 8,000lb capacity low profile Lo-Range trucks were made from the mid 1970s

British Airways were a major user of Ransomes electric tractors.

The pneumatic tyred E-Range fork trucks were suitable for indoor and outdoor use.

Launched in 1974, R-Series reach trucks could operate in the narrow aisles between high warehouse shelving.

The Ransomes Film Unit

Like many other businesses, Ransomes were keen to take advantage of film to promote their products. Their film-maker was Don Chipperfield who joined the company in 1933 as a cost clerk and began to make short films of Ransomes, products from time to time. His early productions were short black-and-white silent films made around the factory to record the manufacture of trolleybuses, ploughs and other equipment.

Following a period of war service, Don returned to Ipswich and set up his own studio in the Orwell Works. As Ransomes' full-time producer he was kept busy scripting and editing or filming on location wherever Ransomes' products were used around the world. He included glimpses of Ipswich in many of his films and he also captured other aspects of Suffolk life.

Don Chipperfield's films were seen by thousands of people benefiting from their free loan to farmers' clubs, agricultural societies and general audiences. Some are still shown by the East Anglian Film Archive and one compilation has been released by them on video and is available from Old Pond Publishing. *Three Films of Ransomes* comprises: "Speed the Plough 1950", showing the Orwell Works and the agricultural implements made there; "Year of Achievement 1961", the foundry and engineering works at the Nacton factory; "Ransomes 1976", new products and new methods at the Nacton Works.

Chapter 11

Machinery for the Export Market

Ransomes' implements were usually given a name when exported but there were occasions when even the Ransomes name was not used on machines sold in some countries. The 1001 combine, for example, was sold in Turkey and Spain as a Ford combine harvester.

Ransomes had their own company or a distributor in most parts of the world and the range of machinery sold in the different countries depended on the local type of farming. Traction engines were sent to Australia and South Africa, thrashing machines to Russia, disc ploughs to Africa and share ploughs to Europe. The use of MG crawlers in French vineyards was extensive although the MG engine suffered power loss at higher altitudes. The first, and not very successful, remedy for this was to fit high-compression heads. Ransomes' engineers then fitted a supercharger to the MGs but this fared little better as it absorbed most of the extra power it produced.

Share Ploughs and Tillage

In the early 1800s Ransomes' animal-draught wooden and iron-framed share ploughs and other tillage implements were exported to various parts of the world including Canada and Australia. Although horses and bullocks were the usual source of power, the *Illustrated London News* for 6 February 1847 reported the use of Ransomes & May wrought-iron and steel elephant-drawn ridging ploughs on Indian sugar plantations.

The V.R.S. improved solid-beam single-furrow iron

Ransomes' elephant plough being used to cut and ridge sugar-cane cultivation trenches five feet apart and deep in proportion. The Illustrated London News *reported that the harness meant that the driver was no longer able to sit on the elephant's neck.*

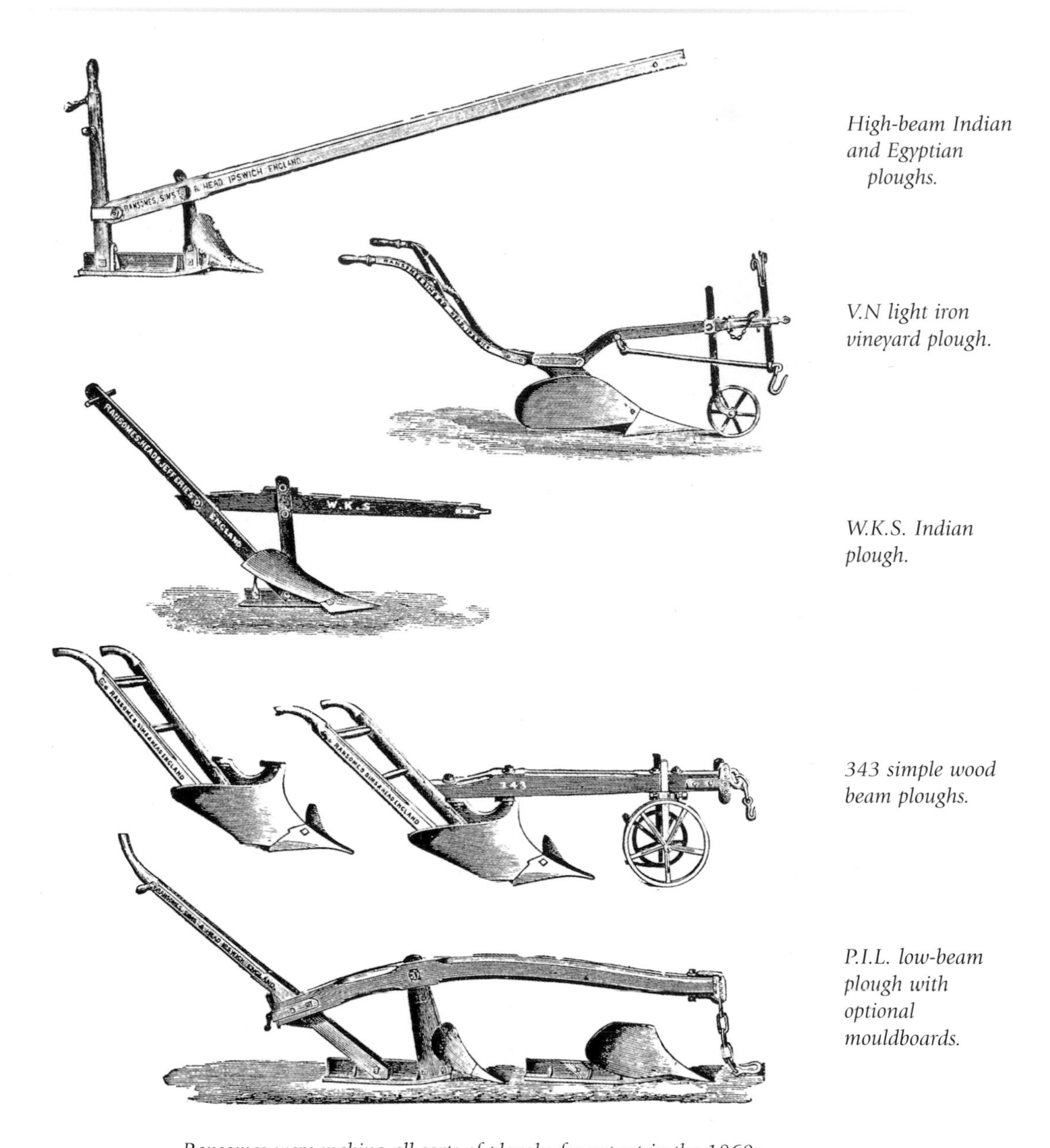

High-beam Indian and Egyptian ploughs.

V.N light iron vineyard plough.

W.K.S. Indian plough.

343 simple wood beam ploughs.

P.I.L. low-beam plough with optional mouldboards.

Ransomes were making all sorts of ploughs for export in the 1860s.

plough, included in Ransomes & Sims' catalogue for 1862, was sold on the home and export markets. Potential customers were advised that V.R.S. ploughs for export were shipped in strong wooden boxes; each case weighed 83cwt, contained six ploughs and required 36 cu ft of cargo space.

Various wooden-beam ploughs for India and tropical countries were included in the 'primitive plough' section in a Ransomes catalogue in 1862. Indian and Egyptian versions of high-beamed ploughs, with or without a single wheel beneath the wooden beam, were said to be a suitable load for a pair of oxen or, with an extended beam, they could be used with an elephant. The Indian and Egyptian ploughs cost £1 10s and £1 15s respectively and were packed five to a box for an extra for 10s. Carriage of course was extra!

The W.K.S. for a pair of small oxen was one of the more basic ploughs for the Indian market. It weighed

The 342D double-furrow version of the 343 wooden-beamed plough with a Ransomes, Sims & Jefferies patent maize drill.

just 27lb with a wooden beam and the K.W.S. of identical design but with an iron beam was twice as heavy. Such similar identification letters combined with communication difficulties must at times have resulted in the wrong ploughs being put in the packing case.

Described as 'simply constructed with as few parts as possible', the wooden-beamed 342, 343 and 344 ploughs for two to four horses could be supplied either complete or alternatively - in order to 'effect a great saving in freight' - customers were able to order the body and handles and have the beam and wheels made locally. The smallest model was also made as a double-furrow plough with a lightweight wooden beam. Designated the 342D for two or three horses and turning furrows 6in deep and 9in wide, it was sold only as a complete plough. The W342M chain-pull wooden-beamed plough for two to six horses could have two, three or four furrows with shares that were raised and lowered into and out of work with a long hand lever.

The P.I.L. plough with a low wooden beam was of primitive design for India and tropical countries. It was supplied with small or large mouldboards, and as with other Ransomes wooden-beamed ploughs, optional small or large ridging bodies were available for an extra 7s and 9s respectively.

French wine producers, who bought considerable numbers of single- and double-furrow V.N. and V.D. light iron vineyard ploughs, were also valued Ransomes customers in the 1860s. Specially constructed for working between rows of vines, the

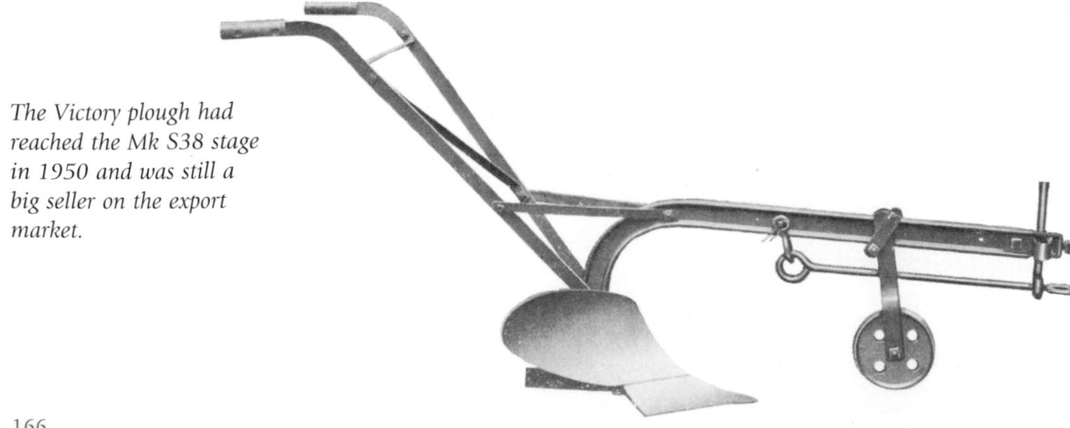

The Victory plough had reached the Mk S38 stage in 1950 and was still a big seller on the export market.

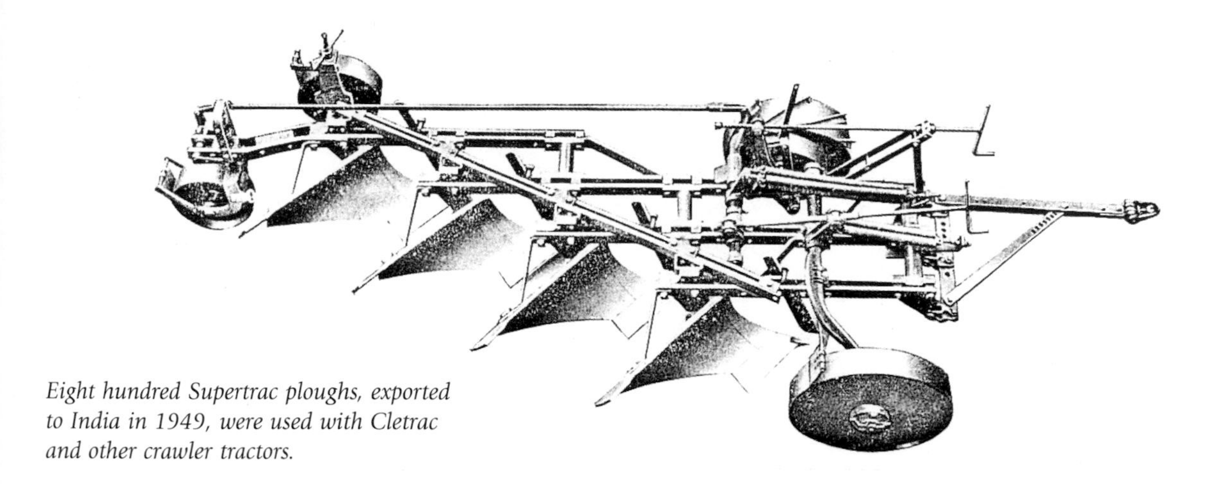

Eight hundred Supertrac ploughs, exported to India in 1949, were used with Cletrac and other crawler tractors.

plough handles had a sideways adjustment allowing the plough to run close to the vines without damaging them.

The single-furrow Victory plough, introduced in 1919 to celebrate the end of the First World War, was designed for use on sugar-cane plantations. The Victory was still made in 1950 when no less than twenty different animal-draught ploughs were listed in Ransomes' 'Cultivation by Animal Draught' catalogue.

Other single-furrow, animal-draught ploughs similar to the Victory and still made in the late 1940s and early 1950s included the 4 to 7in deep and 12in wide furrow E.C.B., five sizes of Cub for up to six horses or oxen and the 'extra strong' No.2 Contractor. This plough, which turned a furrow 12in wide and 4 to 8in deep, was suitable for general ploughing and breaking up new ground. The No.2 Contractor was also recommended for road- and dam-making, which explains why a 1949 sales leaflet for the plough was printed in Dutch as well as English. The animal- or tractor-draught No.2 Elf was a low-profile two- or three-furrow plough for two or four oxen. With an 'absence of projecting parts which might foul trees or bushes', it was an ideal plough for fruit growers.

Most types of Ransomes' tractor share ploughs were sold to farmers throughout the world but some were specially designed for the export market. Arthur Brown, who spent fifty years with the company - many of them as chief plough designer - developed many different models including the Motrac, Multitrac and Supertrac. The Motrac and Multitrac were sold in vast numbers and were used extensively in the reclamation of the Dutch polders. Extra outrigger bodies of the type used on the Victory plough were fitted to help Dutch farmers plough as close as possible to the edge of the dykes.

The Supertrac was specifically developed for the Indian market. Weighing 3½ tons, the four-furrow trailed Supertrac was the largest share plough made by Ransomes. Built at a rate of twenty-five a week, a total of eight hundred Supertracs were shipped to India in 1949. A further batch of fifty ploughs with rubber tyred wheels was also exported, but the tyres suffered considerable damage from large stones.

Disc Harrows

Ransomes' tractor-drawn disc harrows were sold in most parts of the world but animal-draught harrows were also made at the Orwell Works for some export markets. The two-gang and tandem No.4 Countess disc harrow in 6, 8 and 10ft working widths was originally made by J. & F. Howard at Bedford in the 1930s. Primarily for animal draught, the No.4 was sold with a seat for the driver, double head wheel on the drawbar and an adjusting lever for each gang of discs. The Countess Junior two-gang and tandem disc harrows were made in 4, 5 and 6ft working widths. Depending on conditions, a team of eight to sixteen oxen would be required for the 6ft tandem harrow.

Improved versions of the No.3 and No.4 Countess appeared in 1949. The front and rear gangs on the No.3 could be independently adjusted to give different cutting angles and all four gangs were independently adjustable from the seat of the No.4 harrow. The No.3 Countess was basically a tractor-drawn harrow but, with a single- or double-head wheel attachment, still available in 1950, it could be used with a team of

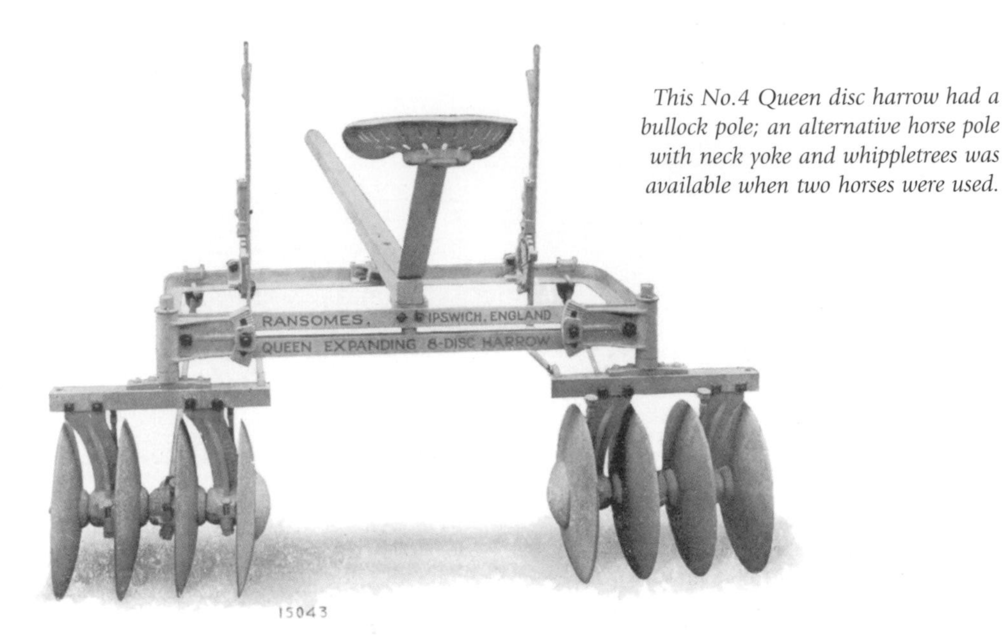

This No.4 Queen disc harrow had a bullock pole; an alternative horse pole with neck yoke and whippletrees was available when two horses were used.

horses or oxen. Sales literature suggested that, when working at sea level with a tractor, the 10ft wide Countess No.3 required about 18 drawbar horsepower. (Internal combustion engines are less efficient at high altitudes.)

Ransomes' Royal and Queen disc harrows, still made in the early 1950s, were designed for working in row crops. The angle of the discs to make them throw soil towards or away from the rows was adjusted with a head lever. The Royal had two sets of four discs carried on adjustable hangers so that the frame and seat could be raised as the crop grew in height. The discs on the Queen were on a fixed frame and a short pole with a hake was provided to hitch it to a tractor. The hake was replaced with a double wheel when pulling the Queen with a pair of horses or oxen and a seat was provided for the driver. By the 1960s Ransomes were selling the same range of HR disc harrows at home and abroad.

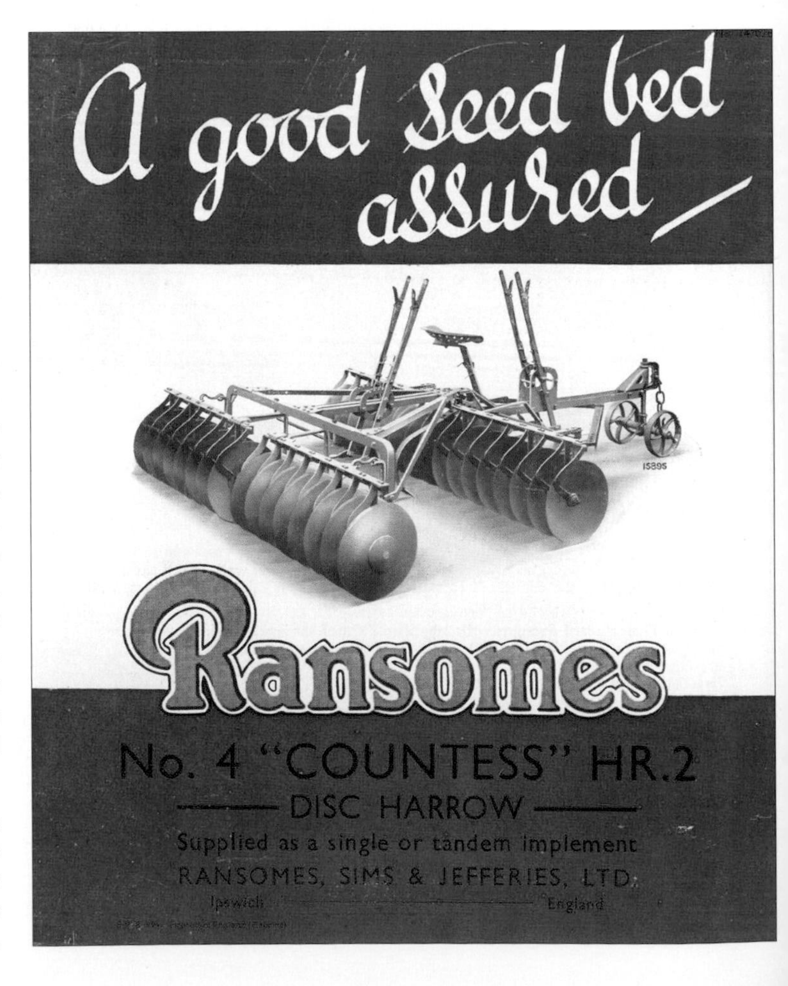

According to soil conditions, a team of ten to twenty oxen was recommended for use with the 10ft wide tandem No.4 Countess HR2 disc harrow and six to ten were required for a two-gang harrow.

Disc Ploughs

The Shugadisc with two, three or four 32in diameter discs cut a 14in wide furrow and could plough-in sugar-cane trash to a depth of 10 to 14in.

Thousands of disc ploughs were built and crated up at the Orwell and Nacton works and exported to many parts of the world including Africa, India, South America and the West Indies. Six types of tractor-drawn disc ploughs with one to eight discs were made in the mid-1930s and in common with other exported implements they had both a name and a model number. Disc plough catalogues at the time included the TD 3 and TD 11 Dragoon, TD 4 Magic, TD 7 and TD 9 Shugadisc, TD 2 Hussar, TD 10 Polydisc, MMS 8 and the Statesman.

Two- and three-furrow Magic reversibles for tractor- and animal-draught were the smallest disc ploughs made by Ransomes in the 1930s. The animal-draught Magics required teams of up to six horses or ten oxen, controlled from a seat at the back of the plough. The tractor-drawn Magic D5 appeared in 1949; the driver controlled the plough from the tractor but the plough seat was retained and when necessary it was used as a rear weight tray. An optional animal-draught pole was available for the Magic D5 plough.

The Hussar and Dragoon, for tractor- or animal-draught, were two- to six-furrow disc ploughs and although the MMS 8 was designed as an eight-furrow model it could also be used with six or seven discs. The massive Shugadisc was designed for ploughing-in sugar-cane trash and the twelve- or sixteen-disc Polydisc could be used with a drill box. Ransomes' animal-draught ploughs in the late 1930s included the Statesman D3 disc plough for a team of eight to twelve oxen.

Some designs of Ransomes disc plough were made

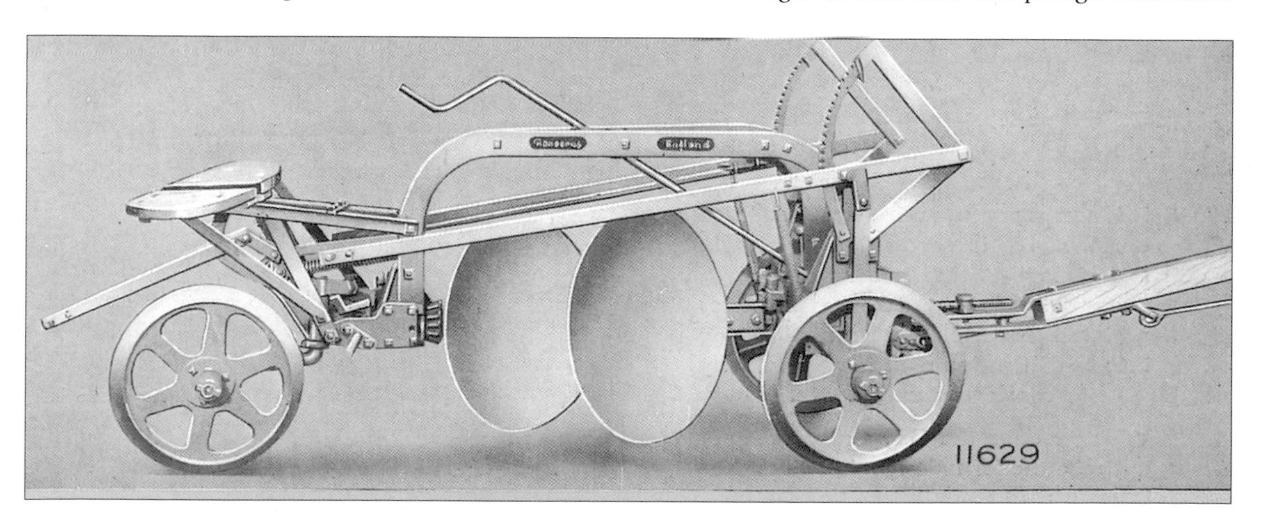

The Magic D 5 reversible disc plough.

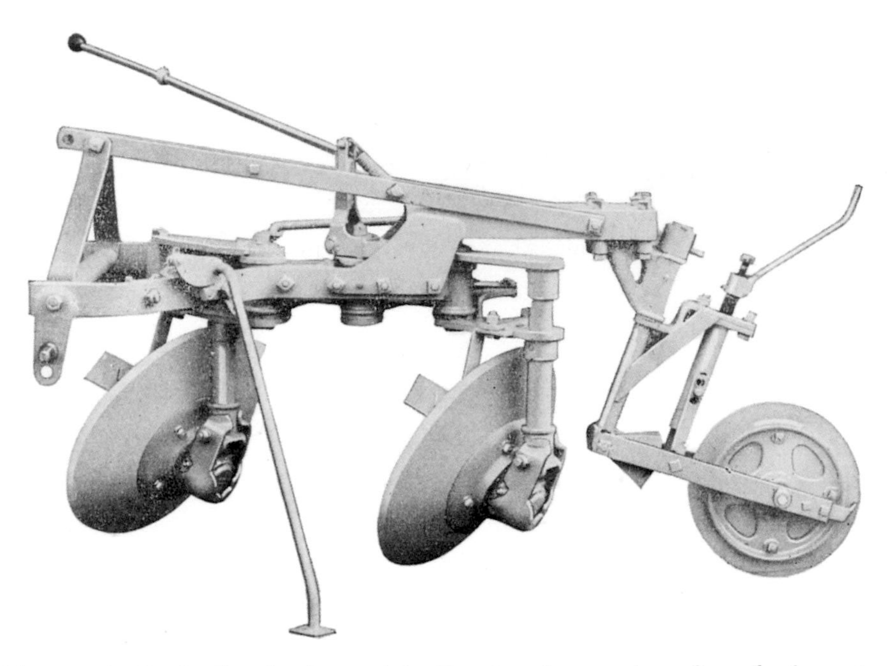

A hand lever on the Condor disc plough turned the discs through an arc in readiness for the next run.

for many years: improved models of the trailed Hussar, Dragoon, Shugadisc and Polydisc, along with the Commando and lightweight MFS were still being made in the early 1950s. Five different TD 12 Hussar ploughs with two to six discs turned furrows up to 12in wide and 9in deep and the two- to six-furrow self-lift TD 11C Dragoon had a maximum working depth of 10in.

Other disc ploughs made at the time included the TD 9A Shugadisc with three or four 32in diameter discs and the twelve- or sixteen-disc TD 10A Polydisc that could be used with a fertiliser and seeding attachment. The lightweight MFS plough with eight 26in diameter discs worked to a depth of 9in and the four- to seven-disc TD 14A Commando for high-powered tractors took furrows up to 12in wide and 14in deep.

New mounted disc ploughs launched in the mid-1950s included the TD 17 / TD 18 Hawk, TD 20 Merlin and the mechanically reversed TD 16 / TD 19 Condor. The Hawk with two, three or four discs and a working depth of 10in was suitable for category I and II linkage and the lighter two- and three-furrow Merlin without a depth wheel was for tractors with a draft-control hydraulic system. The two- or three-furrow Condor reversible disc plough was designed for medium-powered tractors. A hand lever was used to reverse the discs at the end of each run and the furrow width could be altered by changing the angle of the discs. The new seven- and nine-disc TD 27 Polydisc with 4ft 6in and 5ft 9in working widths appeared in 1969 with recommendations that it was the right tool to cultivate around young oil palm and rubber trees and control undergrowth in plantations of citrus fruit, olive, coconut and other trees.

The TD 18 and TD 19 remained in production into the 1970s. By 1983 the somewhat reduced range of Ransomes disc ploughs consisted of the two- or three-furrow TD 17 and four- or five-furrow TD 32 reversibles as well as the TD 30 standard model with four or five discs for 75 to 120hp tractors.

Thrashers and Shellers

Ransomes' thrashing machines designed for European farming conditions were not suitable for all their export markets, but sales literature for the narrow L.A.L. and S.A. thrashers explained that they were well suited to 'travelling on mountainous roads or for short stands of thrashing at distant points'. A 10 to 12hp engine was required for the 36in L.A.L. thrasher while 8 to 10hp was considered sufficient for the S.A. thrasher with a 27in wide drum. The S.A thrasher was called the Pigmy for the export market and, as with all other Ransomes thrashers, the price included a pair of horse shafts or a tractor drawbar, a

The Ransomes S.A Pigmy thrasher with a 27in drum packed down to an overall width of 5ft 3in for transport.

Thrashing in a 'hot country' in the 1880s with a Ransomes thrashing machine and apparatus for chopping and bruising the straw.

The Overseas Representative

The export market was a cornerstone of Ransomes' business. The company had its own offices in Argentina and South Africa. The rest of the world was roughly divided into three sales areas with a farm machinery representative covering South-East Asia and much of Africa, another dealing with Central America and most of South America and a third responsible for parts of Europe.

Perry Crewdson joined Ransomes in 1946 and his forty-two years with the company included spells in the export, publicity and public relations departments. He was one of numerous Ransomes people such Charlie Nice, Fred Dyer, Jimmy Gass, Arthur Brown, Ernie Roworth and Arthur Stephenson, who spent their entire working lives with the company and knew the farm machinery business inside out. His time in the export department was mainly spent in Africa and most countries to the east of West Pakistan. Large quantities of disc ploughs were sold in Kenya, Morocco, Southern Rhodesia, Thailand, West Pakistan and other countries. Trouble-shooting was an important part of an export representative's job. On one occasion Perry found himself staying at a Catholic Mission in the middle of the bush in the Belgian Congo, where he encountered problems with tractor-drawn Dragoon disc ploughs. The Catholic brothers at the mission were experts at carpentry, plumbing and many other trades and it was not unknown for them to hold parties for their visitors which, according to Perry, could get a bit wild at times. On another visit to Kenya Perry saw one of the last teams of sixteen oxen pulling a two-furrow Statesman disc plough. The large team was necessary because oxen do not really pull but just wander along while the driver talks to them non-stop. The driver was assisted by a boy walking in front with a stick over his shoulder attempting to lead the team in something like a straight line. Turning the team on the headland was a major operation!

Ransomes' mouldboard ploughs and field mowers were in demand on the northern-most Japanese island of Hokkaido, where the climate and farming are similar to that in the United Kingdom. While on a sales mission to Japan, Perry Crewdson gained the distinction of being the first Ransomes employee to fly over the North Pole on his way to Tokyo. World travel was more complicated in the early 1960s than it is today, as the Ransomes man discovered when he found himself in the Belgian Congo two days before it became an independent state. Crossing the Congo River in a 12ft dinghy was his only way out of the country en route to Angola, and having got through customs he found that items of his hand luggage had found their way into three different taxis. A Dutch auction was held amongst the taxi drivers at the customs office and the one offering the lowest fare took Perry and his hand luggage to the airport.

Ransomes sometimes developed or modified a machine required by their overseas customers. The rotary slasher and TD 17 mounted disc ploughs were originally developed for farmers in parts of South and

Central America, Asia and Africa. Special bar-point ridging bodies with adjustable mouldboards, similar to those on the Emscott and made for the C68 toolbar, were a top seller in Sudan where they were used to make water channels for an extensive system of gravity irrigation on the Gezira. Ceylon was another important Ransomes customer. Cultivators were used to puddle the rice paddy fields and Ransomes trailers were extensively used to carry produce around the island.

After ten eventful years in Ransomes' export department during which Perry had the satisfaction of seeing crops growing on virgin land ploughed for the first time by Ransomes ploughs, he was concerned with company publicity until retirement in 1989.

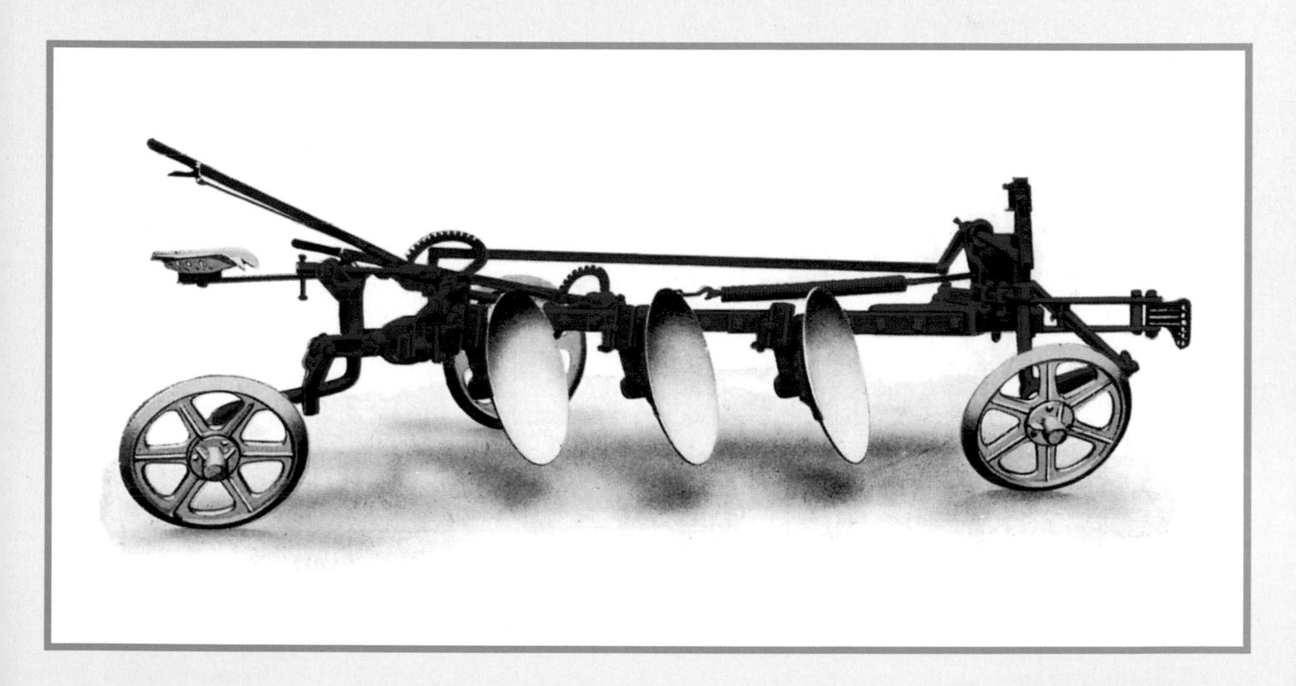

Ransomes' overseas representatives sold hundreds of Statesman D 3 disc ploughs in Africa.

The FR Nacton straw-chopping thrasher was exported to countries where straw was chopped and bruised for animal feed.

full set of tools, a waterproof cover and a long ladder.

A brochure published in the late 1800s explained that it had previously been the custom in some hot regions, such as Egypt and other north African countries, to thrash grain by driving cattle over the sheaves placed on a mud floor. Cattle in those countries were almost entirely fed on wheat straw which was harder and less digestible than the straw of European wheat varieties. However, treading the straw with cattle during the 'thrashing process' softened the fibres and broke them down into a more digestible state. European thrashing machines were not designed to break down the hard wheat straw so Ransomes introduced a straw chopping and bruising apparatus. Straw from the straw walkers was reduced into small pieces and softened 'the same as when the grain was trodden out by cattle. In addition, the straw thus chopped up is entirely free from dust, dirt or dung.'

Although thrashing-machine sales were virtually non-existent in Europe by the early 1950s, Ransomes were still building machines for export. The FR Nacton straw-chopping thrashers were introduced during the period of the agreement between the Ford Motor Co. and Ransomes for the manufacture and sale of FR machinery. The design of the S.Ch 42 and S.Ch 48 machines was said to have been the result of many years of experience with straw-chopping thrashers in Spain, North Africa, Egypt and other regions. Ransomes' thrashers almost invariably had a rasp-bar thrashing cylinder, but the FR Nacton machines had a 3ft 6in or 4ft wide American-type peg drum and on leaving the walkers the straw passed into the chopping apparatus. It was a modern version of the chopping mechanism originally developed in the 1880s for hot countries.

Manufacture of maize shellers was recorded at Orwell Works in 1892. Both hand- and steam-powered shellers were made for the next eighty years or so, originally in the thrasher works and at a later stage by Hunts of Earls Colne.

The Dragon maize sheller, introduced in 1910, was exported to various countries including Argentina and South Africa and considerable numbers of small hand- and pedal-powered shellers, including the Rhino, were also made during the first half of the twentieth century. In the early 1950s Ransomes' maize shellers included the manually operated Demon, the No.4 Hippo (either with a hand crank or 1½hp petrol engine), the Moon belt-driven from a 4

Restored to its original condition, this Dragon maize sheller was photographed in 1999 at an agricultural show in Argentina.

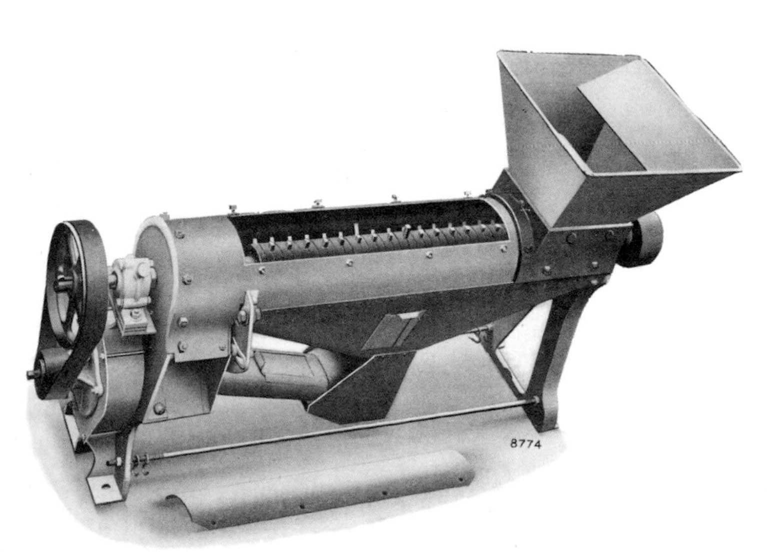

The Moon maize sheller, made for about thirty years, shelled between twenty-eight and fifty 200lb bags of cobs in an hour.

to 6hp engine and a high-output, power-driven model on four wheels.

Hunts of Earls Colne were making the Ransomes Hunt Cobmaster and other small maize shellers in the late 1960s and within a few years all Ransomes' shellers were made in Essex. The power take-off driven Ransomes Hunt Lion maize sheller launched in 1976 was a high-output, tractor-drawn machine. Together with the medium-output, power-driven Puma, Moon and Cobmaster it was produced in the late 1970s, and the Hunt Cobmaster was still being made in 1983.

A portable peanut picker in two sizes was another of Ransomes' export products between the mid-1940s and early 1960s. With a power requirement of 5hp and 10hp respectively the hand-fed Minor and Major

The Ransomes Hunt Cobmaster for the smaller farm could be operated with a hand crank, pedals, electric motor or small petrol engine.

pickers, made by Hunts at Earls Colne, combed the groundnuts from the previously harvested roots with a special design of peg-type thrashing drum and concave. The picked shells were cleaned on a shaking rack and dressing shoe before being bagged up and sent to be washed and bleached for export. A groundnut harvester developed in conjunction with the National Institute of Agricultural Engineering and made by Ransomes was less successful than the peanut pickers, mainly because it was cheaper to harvest the crop by hand than by machine.

Sugar, Tea, Coffee and Cotton

Many standard Ransomes implements including the Baron and Baronet disc harrows, C1 and C59 subsoilers and C17 cultivator/ridger were exported to countries where sugar, tea and coffee were major cash crops. Ransomes' interest in machinery for sugar-cane cultivation prompted them to produce extensive literature on the subject, including a thirty-two page booklet published in 1947. Large crawler tractors were illustrated burying cane trash with TD 9 Shugadisc disc ploughs or TS 41 Supertrac and other share ploughs. Where oxen were the only source of power, the Victory and Cub steel-frame share ploughs and the Sphere all-steel ridging plough were shown working on sugar-cane plantations.

As mentioned previously, the

Ransomes Minor and Major peanut pickers processed up to 800 and 1,100lb of nuts respectively in an hour.

Battery of Giantrac ploughs and Caterpillar tractors on the Waterloo Estate, Trinidad.

single-furrow Victory plough was designed specifically for work on sugar cane. Pulled by one or two horses, oxen or mules, the Victory had a single wheel and a cast-iron mouldboard. During the peak of their production two hundred or so of these ploughs were made every week, thousands of them being built between 1919 and the early 1950s when production finally came to an end.

The Pilot planter, made at the Orwell Works in the 1920s, was used to plant various crops including maize and cotton. It was based in a simple horse-plough frame with a furrower at the front and a seedbox with the feed mechanism driven by a single wheel. An alternative model of the Pilot planter had a small fertiliser hopper and feed mechanism, also driven from the land wheel.

Ransomes' engineers devised a motor cultivator to control weeds growing between rows of tea bushes. The model illustrated was powered by a dry-sump Ransomes petrol engine; a man walking in front was required to steer the cultivator with a long tiller handle while a second man walking behind was in

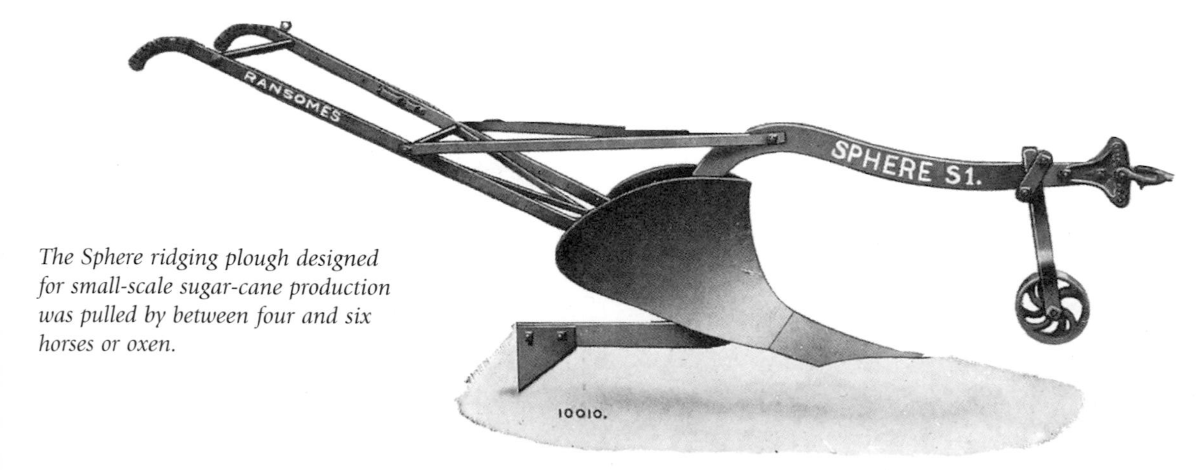

The Sphere ridging plough designed for small-scale sugar-cane production was pulled by between four and six horses or oxen.

This Ransomes motor cultivator with a small set of discs photographed in 1923 was designed for tea plantation work, but it got no further than the prototype stage.

control of the machine. Deflectors guided foliage away from the wheels, and gauntlets on the handlebars were meant to protect the operator's hands from overhanging branches.

Photographs also exist of the self-propelled Hutchinson Motor Tea Cultivator and the Hamilton Motor Plough with a single-cylinder Ransomes petrol engine. Both machines, made in the early 1920s, were controlled with a pair of handlebars by an operator walking behind.

Ransomes were the sole manufacturers of Charles Ansell's patent tea sorting and winnowing machine that had been 'severely tested for five seasons' before a full description of the machine was published in 1886. Numerous testimonials praised its performance in actual work and Mr Ansell's machine had 'received the cordial approval of experts'. It required little power to operate and the tea was thrown into a hopper to be sorted and winnowed in one operation to 'deliver the different qualities of tea with great regularity and quite separate from each other'.

A low-volume, air-blast coffee sprayer similar to an orchard sprayer and claimed to 'herald a complete revolution in coffee spraying' appeared in the early 1950s. The power take-off driven coffee Agro sprayer on a solid chassis mounted on pneumatic tyres was based on the principle of the

A price of £70 was quoted in 1886 for Mr Ansell's patent tea sorting and winnowing machine when packed and delivered to the London docks.

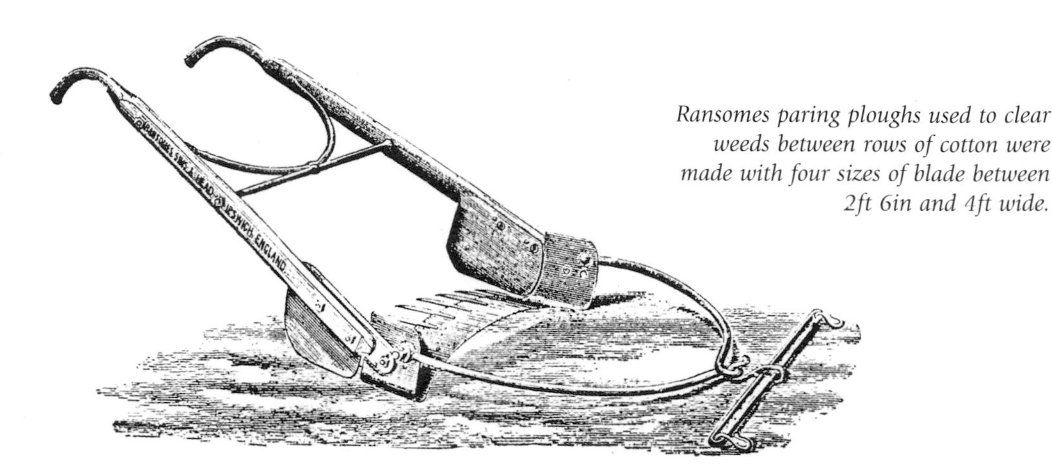

Ransomes paring ploughs used to clear weeds between rows of cotton were made with four sizes of blade between 2ft 6in and 4ft wide.

earlier Ransomes Agro air-blast sprayer for potatoes. It had a 150 gallon tank, centrifugal pump, air blower and an arc of spray jets at low level on both sides of the chassis which applied liquid at 36 gallons an acre. Coffee growers at that time were applying 150 to 200 gallons an acre with existing high-volume machines, but in spite of the money saved in reduced labour costs in carting water, the coffee Agro was not a success.

On the other hand Ransomes did enjoy success in the 1930s and 40s with their Emcot ridger, thousands of which were sold for ridging cotton in Nigeria, Ghana and other countries. The Emcot was similar to the Sphere ridger used by cotton growers and the body could be adjusted to give a distance of 25 to 29in across the back of the mouldboards. Two oxen were needed to pull the Emcot when set at its maximum 12in working depth.

One- and two-horse paring ploughs for cutting weeds between rows of cotton were a much earlier Ransomes contribution to the mechanisation of that crop. Made at the Orwell Works in the 1880s, the paring plough had a straight or curved blade with iron prongs at the back. The blade lifted the grass and weeds between the rows and the prongs let the loose soil fall through, leaving the weeds exposed on the surface to die. Side deflectors on the wooden handles directed the loosened weeds away from the rows of cotton plants.

From the early 1930s to the late 1950s, Ransomes' land levellers were involved in a multitude of projects including irrigation, drainage and road-making. They helped build dams and contour ridges, they levelled ground for golf courses and they were even used in the construction of aerodromes.

The M3, M4 and M5 levellers had a 5, 6 and 7ft wide cutting blade in front of the 'spoil bowl' which, when full, was tripped with a cord from the tractor seat. The bowl rotated to release the spoil and, with a further pull on the cord, the bowl returned to its loading position.

Ransomes land levellers could be set to cut the soil at various depths and then dump the spoil in a heap or spread it in 3 to 7in thick layers.

Chapter 12

Textron

In 1998 Textron Inc, a multi-national corporation with its headquarters in the United States and global interests in the aircraft, automotive, industrial and finance industries, acquired Ransomes plc and its subsidiary Ransomes, Sims & Jefferies Ltd. Ransomes, with its manufacturing facility at Nacton, became part of the Textron industrial division known as Textron Golf, Turf and Specialty Products with responsibility for international distribution of products outside the USA.

Today, Ransomes turf care equipment along with that from Jacobsen, Cushman, Ryan and E-Z-GO, are manufactured and distributed through more than seventy outlets world-wide. The company is still manufacturing and designing new turf care machinery including mounted, trailed and ride-on gang mowers, motor mowers and rotaries mainly for the municipal market. Products manufactured in the United States include Jacobsen ride-on mowers for cutting golf greens, tees and fairways, the Cushman Turf-Truckster maintenance vehicle with its various attachments such as top dressers and core harvesters. Ryan products include turf aerators, de-thatchers and turf cutters and E-Z-GO makes a range of golf cars and industrial vehicles. Textron also distributes the Japanese-built Iseki compact tractors and ride-on rotary mowers in the UK and Ireland, and in 1999 Ransomes introduced a new range of walk-behind and ride-on sweepers.

The Ransomes Commander all-hydraulic ride-on gang mower for cutting parklands and golf fairways has a 51hp diesel engine and a top mowing speed of 7½ mph.

A single pedal controls the forward and reverse hydrostatic transmission on the three-wheel drive, rear-wheel steered Jacobsen triplex greens mower. A range of quick-fit attachments include three fully floating seven-blade cutting cylinders for tees and fairways, three eleven-blade cylinders for greens, turf rollers, spikers and brushes.

Hollow tines on the ride-on Ryan GA-30 Aerator remove cores from the turf to aerate greens, fairways and sports fields at speeds of up to 6mph.

Collecting the cores left by a turf aerator is one of many jobs that can be done with the three- and four-wheel Cushman Turf-Trucksters. With a three-speed automatic or four-speed synchromesh transmission, Turf-Trucksters are also used for haulage, top dressing, spraying and general turf care work.

The diesel-engined Iseki ride-on mower has a stepless forward and reverse hydrostatic transmission and power steering. The grass trimmings are discharged into a high-level dump box from the rear of the cutting deck.

Ransomes Pathway walk-behind and Pathfinder ride-on municipal sweepers, for cleaning shopping precincts, footpaths, and other areas with restricted access, have floating brushes, a vacuum system to collect litter and an optional wander hose. The Pathway sweeper has a water system to suppress dust, and litter is collected in large plastic bags. The water system is optional on the Pathfinder and litter is collected in a high-tip rear dump hopper.

E-Z-GO golf cars were introduced to American golfers in 1954. Today's players have the choice of a golf car with a battery electric motor or a 9hp petrol engine.

Appendix 1

The Ransomes Alphabet

In modern times the letters and numbers used to identify Ransomes' ploughs and plough bodies were sometimes chosen at the whim of the sales manager of the day. It is said that the SCN plough body was named by Scotsman Jim Gass to commemorate his compatriots' theft of the Stone of Scone from Westminster Abbey.

Tractor ploughs were given the 'trac' suffix including the Multitrac, the five-furrow Quintrac, the two-furrow Motrac and the single-furrow Unitrac. The Midtrac fitted conveniently between Motrac and Multitrac. Major and Minor were sometimes used to denote a heavy or light version of a particular plough or other machine. A single letter - A,B,C,D,E,F, etc., - after a model number denoted a modified version of the original implement or fittings for a particular model of tractor. The C18C trailed cultivator, for example, was an improved version of the C18A and C18B.

Many of the letters in the Ransomes farm implement alphabet go back to the days of horse ploughs and the meaning of a letter often depends on its position in the sequence. For example, R used as the first letter denotes Ransomes and as the third or last letter was used for ridger. Although it is not exhaustive the following list includes many of the letters used in the Ransomes alphabet:

A Used for the first ploughs, including the wooden-beamed **AY** made for Norfolk and Suffolk farmers.

Also used as a series letter for an implement to denote either a modification or type of tractor the implement would fit, e.g. TM1021**A** field mower for Fordson Major and Dexta.

AD Argentine deep **D**igger plough body.

B **B**alance plough, and also as a series letter for an implement.

BP **B**ar **P**oint.

C **C**hilled or chilled pattern.

C was also used for cultivators and some mounted toolbar model numbers, e.g. C1 and C8. Cultivators were often given names. Horse-drawn models included the Ipswich, the Orwell and the Small Holdings. The tractor-drawn Dauntless self-lift cultivator carried sequential numbers including No. 3, 9, 10, 11 and 13 to indicate different specifications.

D **D**igger

Used a last or nearly last letter, it signified a **D**ouble-furrow plough.

Also used for **D**isc ploughs, and **TD** was a **T**ractor **D**isc plough.

DC Used in the early 1800s to indicate a **D**ouble **C**hilled mouldboard.

E **E**lephant-drawn plough!

F Indicated the plough was designed for a **F**ordson tractor.

FR **F**ord **R**ansome.

G **G**arden plough.

H **H**arrow. **H**eavy when used as the last letter.

I **I** for **I**rish.

K Used as a first letter for a **K**entish style plough.

L **L**incolnshire or **L**ancashire. **L**ight when used as one of the last letters.

M Used as the last or almost last letter for a **M**ultiple plough (usually three or four furrows).

Also denoted a **M**alleable frame.

Used for **M**otor in the early days of tractor ploughs.

And **TM** tractor field **M**ower.

N Newcastle as a second letter for a type plough.

P For Pattern, Plough or Potato.

PM Plough Mounted.

R Ransomes when the first letter and as the second letter in **HR** (Harrow Ransome).

Ridger when placed later in the sequence.

Reversible plough.

RH Ransomes Horse plough. The RH was followed by a series letter, e.g. R.H.A., R.H.B., etc.

S Scottish when used as a first letter for horse ploughs.

Also denoted Share series ploughs.

Subsoiling plough.

SH Small Holders.

SL Self-Lift.

T Tractor.

Toolbar.

As one of the last letters for Turnover, Turnwrest or Turnabout.

TD Tractor Disc plough.

TN Trailer Non-tipping model.

TS Tractor Share plough.

TSR Tractor Share Reversible - a post 1945 addition.

TT Trailer Tipping model.

V V was for Victory - the animal-draught export plough made in the late 1940s.

W Wooden beam when used as the first letter for a plough.

Wrought-iron beam or frame when used as a last letter.

X Used on some implements made for eXport. Some implements with a model number in the U.K were given a name for the export market.

Y Yorkshire in any position in the sequence.

Some examples of the use of the Ransomes alphabet

IRDCPL Irish Ransomes Double-furrow Chilled Pattern Light plough.

RSLD Ransomes Self-Lift Double plough.

RSLM Ransomes Self-Lift Multiple furrow plough.

SCPB Scottish Chilled Pattern Balance plough.

TSR Tractor Share Reversible plough.

Ransomes also used standard identification letters for bolts, springs and some plough parts.

DH Castings for disc harrows.

PC Castings for ploughs.

RPD Castings for potato diggers.

DR Fabricated components.

PSF Nuts and bolts.

Z Indicated a part was not interchangeable with other parts with the same number

Appendix 2

MG Tractor Serial Numbers

MG 2

1936	101	
1937	143	
1938	267	
1939	398	
1940	533	
1941	730	
1942	934	
1943	1148	
1944	1346	
1945	1652	
1946	1939	
1947	2319	
1948	2675 to 3202	The last MG 2 was made in June 1948.

MG 5

1948	3501	
1949	3590	
1950	4743	
1951	5826	
1952	7238	
1953	8223 to 8652	The last MG 5 was made in June 1953.

MG 6

1953	9001	
1954	9115	
1955	10,095	
1956	10,809	
1957	11,607	
1958	12,500	
1959	13,319	
1960	13,800	The last MG 6 was made in June 1960.

MG 40

1960	13,801	
1961	14,000	
1962	14,501	
1963	14,750	
1964	14,950	
1965	15,170	Production ceased in 1965.

Appendix 3

Company Dates and Names

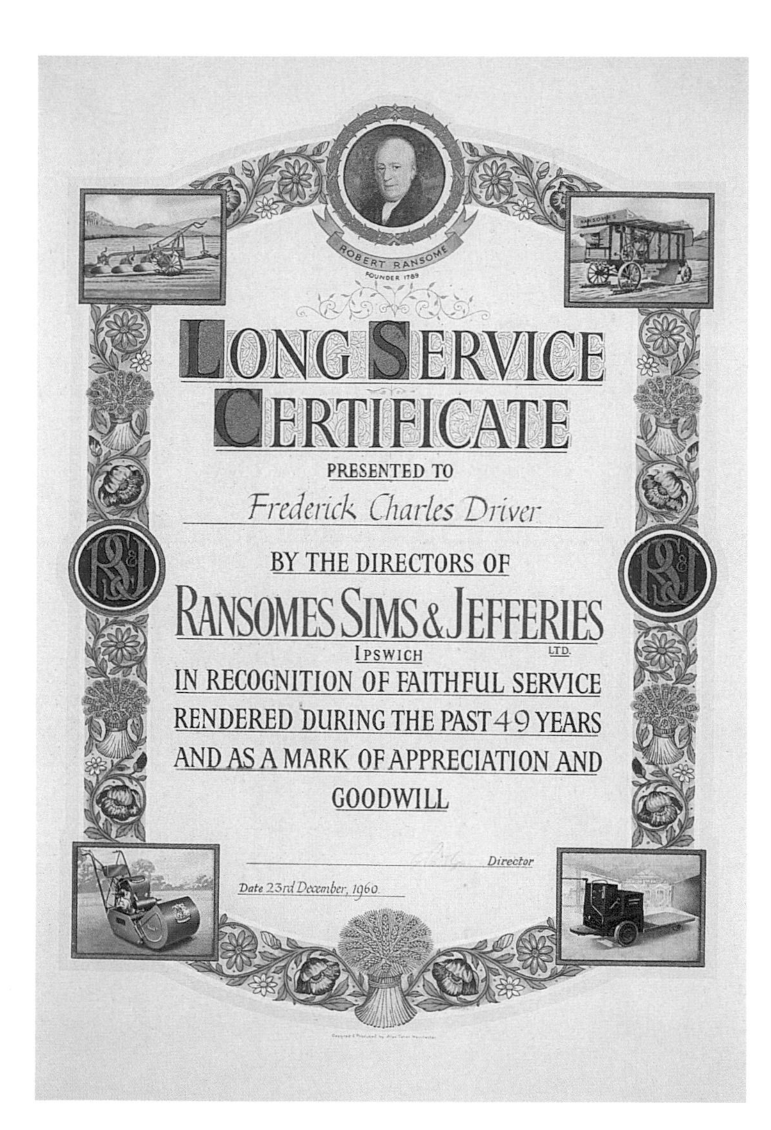

1774 Robert Ransome at Norwich.

1784 Ransomes & Co.

1789 Robert Ransome moved his company to St. Margaret's Ditches, Ipswich.

1809 Ransome & Son.

1818 Ransome & Sons.

1825 J. & R. Ransome.

1830 J. R. & A. Ransome.

1841 Move to Orwell Works began.

1846 Ransomes & May.

1849 Move to Orwell Works completed.

1852 Ransomes & Sims.

1869 Ransomes, Sims & Head.

1869 (Ransomes & Rapier formed).

1881 Ransomes, Head & Jefferies.

1884 Ransomes, Sims & Jefferies.

1911 Ransomes, Sims & Jefferies Ltd.

1949 New foundry opened at Nacton.

1966 Move to Nacton completed.

1968 Orwell Works sold.

1987 Electrolux Group bought farm machinery division and traded under the name Agrolux.

1998 Textron acquired Ransomes plc and became Textron Turf Care & Specialty Products, registered in the UK as Ransomes Jacobsen Ltd.

2001 Name changed to Textron Golf, Turf and Specialty Products.

Index

Books and DVDs from

Ransomes and their Tractor Share Ploughs

Having thoroughly researched the Ransomes archives, Anthony Clare classifies and shows the ploughs produced by the company. He deals with the horse ploughs adapted for tractors, mounted plough development, Ford-Ransomes, the links with Dowdeswells and also includes analyses of identification codes and the TS classification.

Farm Machinery Film Records: Harvest and Rootcrops

Brian Bell has selected footage from the 1940s and 1950s which shows a wide range of farm machinery in development at the time. The films – in black and white – come from the archive of the National Institute of Agricultural Engineering and its Scottish counterpart Brian Bell's informative commentary provides the background to fifty machines, split evenly between two DVDs.

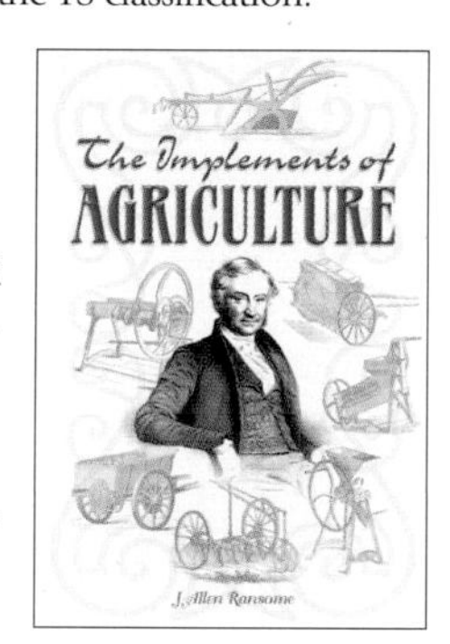

The Implements of Agriculture

Written by J Allen Ransome and originally published in 1843, this generously illustrated book summarised the developments of ploughs, thrashers, drills, milling machines and a wide range of horse-powered implements. 'The best treatise on implements and machinery before the age of steam', The Institute of Agricultural History.

Fifty Years of Farm Tractors

Hardback book by Brian Bell

Dealing with over a hundred companies, in an A-Z format, Brian Bell describes the models and innovations that each contributed to post-war tractor development. With over 300 illustrations, the book is a unique guide to the wide range of machines to be found on Britain's farms.

Seventy Years of Garden Machinery

Covering the period 1920-1990, Brian Bell's book deals with the development of smaller machines: 2-wheeled garden tractors, rotary cultivators, 4-wheeled ride-on tractors, ploughs, drills, cultivators, sprayers, grass-cutting equipment, small trucks, miscellaneous estate items.

Classic Farm Machinery 1

Brian Bell's compilatgion of archive farm machinery films, covering the period 1940-70. Most of the well-known manufacturers are represented, as well as some less familiar names. A broad range of equipment has been chosen to demonstrate developments in machinery for ploughing, cultivating, drilling, spraying and harvesting.

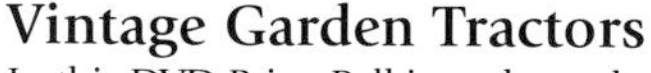

Vintage Garden Tractors

In this DVD Brian Bell introduces the varied and interesting pedestrian-controlled and ride-on garden tractors that took market gardening out of the era of the spade and horse. Featuring fifty models of garden tractor this DVD will appeal to both the novice and the enthusiast. The machines span from the mid-1930s to the mid-60s.

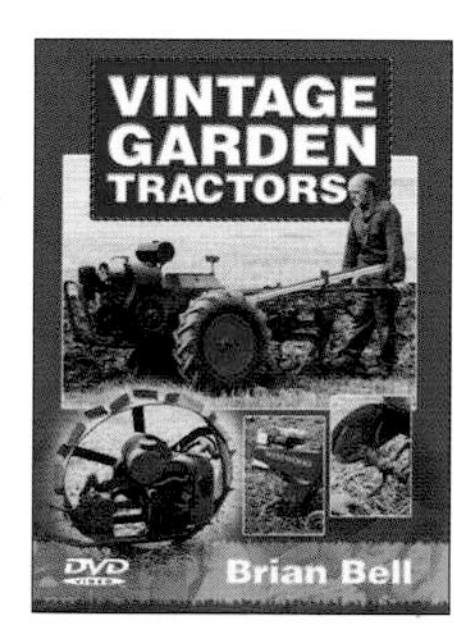

Classic Farm Machinery 2

The second compilation by Brian Bell shows advances in farm mechanisation from the 1970s to the 1990s, using archive footage from most of the major manufacturers. Dealing with equipment for ploughing, cultivating, drilling, spraying and harvesting, the video highlights the emergence of ever more powerful machines with electronic controls.

Enquiries and catalogue:

Old Pond Publishing Ltd, Dencora Business Centre, 36 White House Road , Ipswich, Suffolk IP1 5LT, United Kingdom
Tel: 01473 238200 Fax: 01473 238201 **Website: www.oldpond.com** Email: enquiries@oldpond.com

About the Author

A Norfolk farmer's son, Brian Bell played a key role in developing agricultural education in Suffolk from the 1950s onwards. For many years he was the vice-principal of the Otley Agricultural College, having previously headed the agricultural engineering section. He established the annual 'Power in Action' demonstrations in which the latest farm machinery is put through its paces and he campaigned vigorously for improved farm safety, serving for many years on the Suffolk Farm Safety Committee. He is secretary of the Suffolk Farm Machinery Club. In 1993 he retired from Otley College and was created a Member of the Order of the British Empire for his services to agriculture. He is secretary and past chairman of the East Anglian branch of the Institution of Agricultural Engineers.

Brian's writing career began in 1963 with the publication of *Farm Machinery* and *Farm Tractors* in Cassell's 'Farm Books' series. In 1979 Farming Press published a new version of *Farm Machinery*, which is now in its fourth enlarged edition, with more than 25,000 copies sold. Brian's involvement with videos began in 1995 when he compiled and scripted *Classic Farm Machinery*.

Brian Bell writes on farm machinery past and present for *Farmers Guide, Tractor & Machinery* and other specialist magazines. Brian lives in Suffolk with his wife Ivy. They have three sons.

Books and Videos by Brian Bell

Books

Farm Machinery, Farm Tractors, Farmyard Machinery, Farm Workshop, 1966 (out of print)
Farm Machinery, 1979, 4th rev. edn 1999
Farm Workshop, 1981, 2nd edn 1992
Machinery for Horticulture (with Stewart Cousins), 1992, 2nd edn 1997
Fifty Years of Farm Machinery, 1993
Fifty Years of Garden Machinery, 1995
Fifty Years of Farm Tractors from Allis-Chalmers to Zetor, 1999.

Videos

Classic Farm Machinery Vol. 1 1940-1970, 1995
Classic Farm Machinery Vol. 2 1970-1995, 1995
Classic Tractors, 1996
Classic Combines, 1996
Ploughs and Ploughing Techniques, 1997
Harvest from Sickle to Satellite, 1998
Power of the Past, 1998
Acres of Change, 1999
Steam at Strumpshaw, 1999
Tracks Across the Field, 1999
Vintage Match Ploughing Skills, 2000
Thatcher's Harvest, 2000.

RANSOME
THE BEST
THE "AUTOMATON